SpringerBriefs on Pioneers in Science and Practice

Texts and Protocols

Volume 23

Series editor

Hans Günter Brauch, Mosbach, Germany

For further volumes:
http://www.springer.com/series/11446
http://www.afes-press-books.de/html/SpringerBriefs_PSP.htm
http://www.afes-press-books.de/html/SpringerBriefs_PSP_TP.htm

Michael Drieschner
Editor

Carl Friedrich von Weizsäcker: Major Texts in Philosophy

Springer

VDW

Knowledge and Responsibility
Carl Friedrich von Weizsäcker Society & Foundation

UDO KELLER STIFTUNG
FORUM HUMANUM

RUB

Editor
Michael Drieschner
Munich
Germany

ISSN 2194-3125 ISSN 2194-3133 (electronic)
ISBN 978-3-319-03670-0 ISBN 978-3-319-03671-7 (eBook)
DOI 10.1007/978-3-319-03671-7
Springer Cham Heidelberg New York Dordrecht London

Library of Congress Control Number: 2014937683

Copyediting: PD Dr. Hans Günter Brauch, AFES-PRESS e.V., Mosbach, Germany

Printed on acid-free paper

Springer is part of Springer Science+Business Media (www.springer.com)

Carl Friedrich von Weizsäcker in Leipzig (1936).

Other Books on Carl Friedrich von Weizsäcker Published in this Book Series by Springer

- Ulrich Bartosch (Ed.): *Carl Friedrich von Weizsäcker: Pioneer of Physics, Philosophy, Religion, Politics and Peace Research.* Springer Briefs on Pioneers in Science and Practice No. 21. (Cham–Heidelberg–Dordrecht–London–New York: Springer-Verlag, 2015).
- Michael Drieschner (Ed.): *Carl Friedrich von Weizsäcker: Major Texts in Physics.* Springer Briefs on Pioneers in Science and Practice No. 22. Subseries Texts and Protocols No. 10 (Cham–Heidelberg–Dordrecht–London–New York: Springer-Verlag, 2014).
- Konrad Raiser (Ed.): *Carl Friedrich von Weizsäcker: Major Texts in Religion.* Springer Briefs on Pioneers in Science and Practice No. 24. Subseries Texts and Protocols No. 12 (Cham–Heidelberg–Dordrecht–London–New York: Springer-Verlag, 2014).
- Ulrich Bartosch (Ed.): *Carl Friedrich von Weizsäcker: Major Texts on Politics and Peace Research.* Springer Briefs on Pioneers in Science and Practice No. 25. Subseries Texts and Protocols No. 13 (Cham–Heidelberg–Dordrecht–London–New York: Springer-Verlag, 2015).
- See also the website on these five books with texts by Carl Friedrich von Weizsäcker, at: <http://afes-press-books.de/html/SpringerBriefs_PSP_C.F.v._Weizsaecker.htm>.

Carl Friedrich von Weizsäcker (1940). © The Weizsäcker Family represented by Dr. Elisabeth Raiser who granted permission to use this photo

Acknowledgments and Sources

The editors express appreciation to the Carl Friedrich von Weizsäcker Foundation and to Dr. Bruno Redeker for the permission to use the texts and translations from the volumes *The Ambivalence of Progress* (New York: Paragon House, 1988) and *The Unity of Nature* (New York: Farrar-Strauss-Giroux - Macmillan, 1980). Both publishers do not list these books any more in their programme nor do they list Carl Friedrich von Weizsäcker as an author. They were informed by Dr. Redeker as the representative of the bequest of Carl Friedrich von Weizsäcker that the copyrights for all publications of Carl Friedrich von Weizsäcker were transferred to the Carl Friedrich von Weizsäcker Foundation. Neither publisher replied to the question whether they plan new editions of the books. Therefore, the rights were retransferred to the Carl Friedrich von Weizsäcker Foundation, which granted permission to reprint all texts.

The financial support of the Udo Keller Foundation is also gratefully acknowledged. The editor further thanks Mr. Mike Headon for language editing and Mr. Thomas Bast for IT support, as well as all colleagues of Springer publishers in Heidelberg and Chennai for their efforts to produce a good book. Special thanks go to Ms. Ann Hentschel for the new translations and, last but not least, to Adj. Prof. Hans Günter Brauch, the editor of the series, for his inestimable constant support.

The texts of Carl Friedrich von Weizsäcker in this volume were initially published in:

1. "Preliminary Epistemological Considerations", in: *Zeit und Wissen* (Munich: Hanser, 1992); Part I, Second Chapter; new translation.
2. "A Description of Physics": in: *The Unity of Nature* (New York: Farrar Straus Giroux, 1980); translation of: *Die Einheit der Natur* (Munich: Hanser, 1971): I, 6.
3. "Time–Physics–Metaphysics", in: *Der Mensch in seiner Geschichte* (Munich: Hanser, 1991), new translation.
4. "Biological Preliminaries to Logic", in: *The Ambivalence of Progress* (New York: Paragon House, 1988); translation of: *Der Garten des Menschlichen* (Munich: Hanser, 1977): II, 6.
5. "Models of Health and Illness, Good and Evil, Truth and Falseness", in: *The Unity of Nature* (New York: Farrar Straus Giroux, 1980); translation of: *Die Einheit der Natur*, (Munich: Hanser, 1971): III, 4.

6. "Parmenides and the Greylag Goose", in: *The Unity of Nature* (New York: Farrar Straus Giroux, 1980); translation of: *Die Einheit der Natur* (Munich: Hanser, 1971): IV, 5.
7. "Parmenides and Quantum Theory", in: *The Unity of Nature* (New York: Farrar Straus Giroux, 1980); translation of: *Die Einheit der Natur* (Munich: Hanser, 1971): IV, 6.
8. "Possibility and Movement: A Note on Aristotelian Physics", in: *The Unity of Nature* (New York: Farrar Straus Giroux, 1980); translation of: *Die Einheit der Natur* (Munich: Hanser, 1971): IV, 4.
9. "The Rationality of Emotions", in: *The Ambivalence of Progress* (New York: Paragon House, 1988); translation of: *Der Garten des Menschlichen* (Munich: Hanser, 1977): II, 4.
10. "On Power", in: *The Ambivalence of Progress* (New York: Paragon House, 1988), translation of: *Der Garten des Menschlichen* (Munich: Hanser, 1977): II, 5.

Contents

1 Carl Friedrich von Weizsäcker: Major Texts in Philosophy—A Brief Introduction by the Editor 1
References 10

2 Preliminary Epistemological Considerations 11
2.1 Similes for the Structure of Physics 11
2.2 Cognition 14
2.3 Doubt 16
2.4 Faith 20
2.5 Methodological Consequences 23
Reference 27

3 A Description of Physics 29
3.1 The Role of Physics 29
3.2 An Anecdote as an Eye Catcher 30
3.3 A Historical Example 32
3.4 Philosophy of Physics 33
3.4.1 Theological Platonism 33
3.4.2 The World as Mechanism 35
3.4.3 Empiricism as a Description of Physics 36
3.4.4 Difficulties of Fundamental Empiricism 38
3.4.5 Scientific Perception 41
References 43

4 Time: Physics—Metaphysics 45
4.1 Metaphysics 45
4.2 Physics 49
4.3 Evolution 52
4.4 Culture 55

5 Biological Preliminaries to Logic 61
5.1 The Question 61
5.2 The Place of Logic 63
5.3 Practical Interpretation of the Bivalence of Logic 64
5.4 The Basis of the Subject-Predicate Structure of Sentences 68
Reference 73

6 Models of Health and Illness, Good and Evil, Truth and Falseness 75
6.1 Health 76
6.2 Illness 80
6.3 Progress 82
6.4 Good and Evil 84
6.5 Truth 86
References 89

7 Parmenides and the Graylag Goose 91
7.1 What This Essay Is About 92
7.2 The Idea of the Graylag Goose 96
7.3 The Form in Things and as Thought 98
7.4 Form as Example 105
References 108

8 Parmenides and Quantum Theory 109
8.1 What Does the Unity of Nature Mean? 109
8.2 A Digression: How Can One Read the Philosophers? 112
8.3 What Were Parmenides and Plato Talking About? 115
8.4 The First Hypothesis of the Platonic Parmenides and Quantum Theory 119
8.5 The Approach to the Second Hypothesis 125
References 126

9 Possibility and Movement: A Note on Aristotelian Physics 127
References 137

10 The Rationality of Emotions 139
10.1 The Concept of Interest 139
10.2 Place and Action (Possession, Domination, Power) 141
10.3 Love 144
10.4 Insight 146
10.5 Values 150
10.6 The Enlightenment 153
10.7 World Peace and Self-Realization 154

11 On Power 159
11.1 The School of Political Realism 159
11.2 Biological Anthropology 162
11.2.1 Evolutionist Thinking 162
11.2.2 The Structure of Instincts in Humans and Animals . . . 163
11.2.3 Stabilization in Inequality Through Selection 165
11.2.4 What Is Power? 166

Carl Friedrich von Weizsäcker Society 171

Carl Friedrich von Weizsäcker Foundation 173

Federation of German Scientists 175

Udo Keller Stiftung Forum Humanum 177

Ruhr University Bochum 179

About the Author 181

About the Editor 185

About the Book 187

Carl Friedrich von Weizsäcker at the age of 87 in 1999. © Lili Bansa who granted permission to use this photo

Chapter 1
Carl Friedrich von Weizsäcker: Major Texts in Philosophy—A Brief Introduction by the Editor

This volume is part of a short series which will make major texts by the German scientist and philosopher Carl Friedrich von Weizsäcker (1912–2007) available in English to a wider audience. Through his writings, many of which have already appeared in English translation, he has become known as a pre-eminent theoretical physicist as well as one of the most important German philosophers of the twentieth century. This volume is devoted to texts in philosophy while the four other volumes in the series present an overview of his writings and selections of texts on physics, religion, and politics and peace research.

Though Weizsäcker's writings are between 20 and 80 years old, progress in philosophy is much slower than in the sciences, and so his writings on philosophy are as modern today as they were when they were first conceived. Besides modern science, his philosophical writings focus on Plato, Aristotle, and Kant; they can help us understand quantum theory, and quantum theory, if it is analysed as Weizsäcker does, can help us understand classical philosophy. It is interesting that the fundamental problems of quantum mechanics are discussed today in almost the same way as they were in the nineteen-thirties, partly because the problems are so difficult, and partly because some of the complex dialogue has been forgotten, so that the same arguments are taken up over and over again.

M. Drieschner (ed.), *Carl Friedrich von Weizsäcker: Major Texts in Philosophy*,
SpringerBriefs on Pioneers in Science and Practice 23,
DOI: 10.1007/978-3-319-03671-7_1,

Weizsäcker's contributions to the discussion of classical philosophy as well as of the fundamental questions of modern science are in any case worth studying. This selection is intended to promote the study of Weizsäcker's thinking, which has not been much noticed in the American discourse so far. This may be, at least partly, because his thinking does not fit well into the mainstream of analytical philosophy and empiricist thought. Speaking of the beginning of his academic studies, Weizsäcker writes[1]

> When I discovered, in my final year of 'gymnasium',[2] that the field that attracted me most was philosophy, I was tempted to study that. He [Heisenberg] advised me that to practise philosophy relevant to the twentieth century one had to know physics, that one learns physics only by practising it, and that one does best in physics before the age of thirty and in philosophy after the age of fifty. I followed his advice, studied theoretical physics, and have never regretted it.

So Weizsäcker was educated as a physicist, but had studied physics because of his interest in philosophy. He unites uniquely within himself interests and abilities in science as well as in the humanities. Of his beginnings as a researcher, he writes:

> that I studied physics in order to understand quantum theory philosophically was obvious. … the philosophical problems of physics did occupy the minds of physicists, but I learned, step by step, that neither they nor contemporary philosophers could get a grip on them. For twenty-five years I escaped into unphilosophical, concrete physics, while continuing to meditate quietly on quantum theory and to depress myself with self-reproach. But physics, like good bread, was nourishing and invigorating.

His account of "unphilosophical physics" is an exaggeration: as early as 1943 a collection of five philosophical papers had appeared under the title *Zum Weltbild der Physik*.[3] This collection was revised and enlarged several times up to 1958, when it contained eighteen papers; the latest edition appeared in 2002.

---

[1] Cf. his "Self-Portrait", in: von Weizsäcker (1988).

[2] The secondary school in Germany.

[3] English translation by Grene (1952).

Essays on
Historical
Anthropology
Carl Friedrich
von Weizsäcker
The
Ambivalence
of Progress

After the war he turned more and more—as he had planned at the beginning of his studies—to philosophical questions. In 1946 he gave a series of lectures in Göttingen on *The History of Nature*[4] that dealt formally with cosmology and evolution, but turned out to become one of his most important philosophical texts. In 1948 another series of lectures followed, about the conceptual structure of theoretical physics; it was not published until 2004.

The texts reprinted here were originally all written in German and translated into English. Weizsäcker was a master of the German language. For this very reason his texts are not easy to translate, since English usage and structure are very different from German usage and structure. Moreover, his style of academic writing belongs to a past era, such that even students of today who are native speakers of German would find it difficult. Weizsäcker's collection *The Unity of Nature* had been fortunately translated by Francis J. Zucker, a Vienna-born scientist who spent his life from the age of 18 in the United States. Later he worked in Germany for more than 10 years as Weizsäcker's collaborator and so knew his writings rather well. Moreover, he was married to an American writer, Ilona Karmel, who helped him a lot in his translation work; all in all, the best conditions for good translation. Apparently the (anonymous) translator of "The Ambivalence of Progress" was much less fortunate. His translation contains many misunderstandings and omissions that required much work for inclusion in this collection. Texts not previously published in English have been newly and very finely translated by Ann Hentschel.

The first text we reprint is the introduction to Weizsäcker's lectures on the conceptual structure of theoretical physics, "Preliminary Epistemological Considerations" (Chap. 2),[5] which he considered important enough to include in his magnum opus of 1992.[6] It can also serve as an introduction to this volume.

"A Description of Physics" (Chap. 3) is taken from a series of lectures in 1968, "Critique of the Sciences". Here Weizsäcker deals with the role of physics in the universe of the sciences (and humanities). The text also provides a critical assessment of empiricism, following the 'Humean challenge' famously reformulated by Karl Popper. But he contradicts Popper in his conclusion that scientific claims can be falsified: not only does any falsification presuppose the truth of other claims that cannot be verified empirically, but even the empiricist prejudice rests on shaky ground. Weizsäcker stresses that there is a special manner of scientific perception, which does not provide certain knowledge but which is indispensable for science.

---

[4] *Die Geschichte der Natur* (Göttingen: Vandenhoeck & Ruprecht, 1948); English translation: *The History of Nature* (Chicago, Chicago University Press, 1949).

[5] Numbers in () refer to the chapters in this volume.

[6] *Zeit und Wissen* [Time and Knowledge] (Munich: Hanser, 1992).

"Time, Physics, Metaphysics" (Chap. 4) is a speech Weizsäcker made in 1983. He introduced it with the remark: "I have been led into temptation and have succumbed to the temptation to sketch the complete outline of a philosophy in somewhat over an hour." We can read it, consequently, as a sketch of what he considers as the main points of his philosophy. One sees in this text the emphasis he lays on classical philosophy, beginning with Plato (his favourite philosopher) and leading up to Kant and Hegel, as well as on the structure of time as present, past, and future. We can also see the special point he makes very often that man is a product of natural history, and that science in history is a product of man that serves, in turn, to observe natural history: there is a kind of circular relationship that does not conform to the deductive structure we usually associate with philosophical reasoning. And one sees the intimate connection of Weizsäcker's philosophy with modern science: he uses classical philosophical thought to treat a fundamental problem of modern science, while on the other hand he sheds new light on ancient philosophical problems by considering the insights of modern science.

Carl Friedrich von Weizsäcker

THE UNITY OF NATURE

"Biological Preliminaries to Logic" (Chap. 5) can be read as an application of the previous text. The question is where logic comes from. Logic deals with truth and falsehood, so it cannot be justified psychologically—as Husserl had already stressed. It is dealt with mathematically, so one might suppose that it depends systematically on mathematics; but that seems to be impossible since mathematics, on the other hand, clearly depends on logic. Weizsäcker views the problem from the perspective of science: what is truth in a biological framework? How might some forerunner of logic have served animals in the course of evolution? How could the truth of logical thinking be tied together with the brain functions that biology describes?

Carl Friedrich von Weizsäcker in the nineteen-sixties. © The Weizsäcker Family represented by Dr. Elisabeth Raiser who granted permission to use this photo

"Models of Health and Illness, Good and Evil, Truth and Falseness" (Chap. 6) can again be seen as a continuation of the previous text. It starts from the question of what health is and tries to provide an answer in the framework of biological cybernetics: in some sense being healthy means conforming to a *norm*, fulfilling the right *functions*, the organism serving its *purpose*. But what does that mean? A Darwinist model, says Weizsäcker, gives precise answers. He compares the ecological niche with Plato's *eidos*, 'Form', where the Form of, for example, a greylag goose (the animal Konrad Lorenz had studied most) is neither the form of a concrete goose nor the class of all greylags. And illness is not only lack of health but a deviation from health that has its own *eidos*—as, for example, in the case of measles. Maybe one could call it 'wrong health'. So the English language provides a fortunate connection between the three pairs, in that the three negative sides may all be called "wrong" in a certain sense: illness is wrong health; doing evil is doing wrong; and an answer that is not true is wrong. Weizsäcker subsumes all three pairs under the aspect of a cybernetic, Darwinian model.

"Parmenides and the Greylag Goose" (Chap. 7) takes up the question of the ecological niche, but this time under the aspect of Plato's 'Parmenides' dialogue. Whereas the previous text tried to elucidate Platonic philosophy using modern science, this text attempts the opposite, elucidating modern science using Platonic philosophy. The 'Parmenides' dialogue is, Weizsäcker says, "a Platonic dialogue for the modern scientist". He gives a running interpretation of the dialogue, "treating it strictly in accordance with the viewpoint of contemporary science". On the other hand, Weizsäcker tries not to impose our contemporary prejudices on the ancient text. He claims that a good guide to understanding Plato is the "principle of the truth of what is asserted": In his dialogues, Plato often makes an interlocutor immediately refute what the speaker has said. But what is refuted is a certain interpretation of what had been said; Plato must have had in mind to thus introduce another interpretation he considered the right one. One of the subjects Plato treats (and Weizsäcker interprets) is the question of where Forms exist—in things or in thoughts? Plato's solution is—very crudely put—that the Form is thought because thinking thinks *something*; what can be thought are the Forms. One of the most interesting topics of the discussion is the relationship between Forms and examples.

"Parmenides and Quantum Theory" (Chap. 8) is again about Platonic dialogue, this time directly confronting its content with quantum mechanics. The previous text dealt with the "Forms of things"; this text now deals with the next step (according to Plato), the Form of Forms, the One. So this is very abstract theory, in physics as well as in philosophy. After a couple of somewhat introductory paragraphs: "What does the unity of nature mean?", "A digression: how can one read the philosophers?" (a general description of Weizsäcker's approach), and "What were Parmenides and Plato talking about?", there follow two paragraphs about the relationship between the Platonic Parmenides and modern abstract quantum theory.

Carl Friedrich von Weizsäcker (Conference on Quantum Logic, Cologne 1984). © Hans Berner who granted permission for its use in this volume

"Possibility and Movement" (Chap. 9) is a reflection on Aristotle's theory of the continuum, confronted again with its modern treatment. Essentially, Aristotle treats the continuum as a problem of physics, not of pure mathematics, and Weizsäcker agrees with Aristotle on this point. The continuum is intimately connected with possibility, a central term of Aristotelian philosophy. Weizsäcker treats the Aristotelian pair 'possibility' or 'potentiality' (δύναμις) and 'reality' or 'actuality' (ἐντελέχεια) in direct relationship with time, where possibility is related with the future. So he draws a line between movement, which according to Aristotle depends on potentiality and in the modern view on space and time, and the continuum, which according to Aristotle depends again on potentiality but is, according to the modern view, a subject of pure mathematics; and both are related to the structure of time as present, past, and future. In this text Weizsäcker refers to Aristotelian terms in Greek and writes them from time to time in Greek letters. But even if the reader does not read Greek he or she will easily recognize these terms after a while; so we have refrained from providing a transliteration in every instance.

"The Rationality of Emotions" (Chap. 10) ties philosophy not to physics but to biology. The question is the motivation for political action: is the motive rational or irrational? What does this distinction mean? What are the driving forces behind politics?—Weizsäcker treats these questions under the subheadings *A. The Concept of Interest*, *B. Place and Action* (*Possession*, *Domination*, *Power*), *C. Love*, *D. Insight*, *E. Values*, *F. The Enlightenment*, and *G. World Peace and*

Self-Realization. The paper was not originally meant for the public, but rather as an internal reflection for consideration in the Max Planck Institute on foreign politics and peace in the time of the Cold War. It is a good example of Weizsäcker's political philosophy.

In the concluding text, "On Power" (Chap. 11), Weizsäcker deals with Political Realism (Niebuhr, Morgenthau, Kindermann). He discusses it at first on a rather abstract level, confronting it with biological anthropology: instinct, inequality through selection, and so on. Finally he asks "What is power?" in the framework of his abstract analysis, relating it to the scientific concepts of energy and information. He concludes that "Power appears as the unavoidable result of the coincidence of any competitive situation and intellectual understanding. In view of the possibility of reason and peace, power is not necessarily the last word. That is for history to decide."

This collection of texts cannot be more than an appetizer for Weizsäcker's philosophical writings. Anyone stimulated to really find out about Weizsäcker's philosophy is invited to read more of his texts. In English, mainly the lecture "The History of Nature", as well as the collections "The Unity of Nature" and "The Ambivalence of Progress" can be recommended.[7]

References

von Weizsäcker, Carl Friedrich, 1952: *The World View of Physics* (London: Routledge and Kegan Paul), translation by Marjorie Grene of 'Zum Weltbild der Physik'.

von Weizsäcker, Carl Friedrich, 1988: *The Ambivalence of Progress* (New York: Paragon House): 1–30.

[7] Cf. the bibliography in the chapter "On the Author" at the end of this volume. A comprehensive bibliography of works mainly in German has been published in: *Journal for General Philosophy of Science.* 2008, Vol. 39, pp. 179–244. DOI: 10.1007/s10838-008-9069-6. See also the website on this book on Carl Friedrich von Weizsäcker, at:<http://afes-press-books.de/html/SpringerBriefs_PSP_C.F.v._Weizsaecker.htm>.

Chapter 2
Preliminary Epistemological Considerations

This chapter originates from the introduction to a lecture course on "The Conceptual Structure of Theoretical Physics" that I held at Göttingen during the summer semester of 1948.[1,2] The topic of that lecture was already basically the one of the present work. The addressees of that lecture were primarily scientists, particularly physicists. The introductory sections reprinted here essentially express, in a casual style, the same basic approach to the 'phenomenal premises' of *The Structure of Physics* (2006), upon which the present book also relies. Hence they can—as I hope—serve as an easily accessible introduction into this way of thinking. More precise details will appear in subsequent chapters in treating the pertinent problems.

In basic outlook the thoughts developed here otherwise also date back to a note from 1932 that I had published in excerpt in the book *Wahrnehmung der Neuzeit*.[3]

2.1 Similes for the Structure of Physics

This lecture is divided up into three parts.

The first part deals with "Elementary Givens". Under these I would like to comprehend everything that is held to be valid methodologically and conceptually as a general precondition for science, particularly physics. The path that leads to general concepts, such as, thing, space, time and generality, starts out from phenomena. Phenomenology is the endeavour to reflect on the givenness of

---

[1] This text was first published in: Carl Friedrich von Weizsäcker: *Zeit und Wissen* [literally: 'Time and Knowledge'] (Munich: Hanser, 1992), pp. 35–59. It was translated for this volume by Ms. Ann Hentschel with the financial support of the Udo Keller Foundation—*Editor* [MD].

[2] von Weizsäcker (2004)—*Editor* [MD].

[3] Carl Friedrich von Weizsäcker: *Wahrnehmung der Neuzeit* ['Perception of Modernity'] (Munich: Hanser, 1983). Cf. there the sections *Bohr und Heisenberg, eine Erinnerung aus dem Jahr 1932* ['Bohr and Heisenberg. A Recollection from the Year 1932'], pp. 134–146, and *Begriffe: Bewußtsein als unbewußter Akt* ['Concepts: Consciousness as an Unconscious Act'], pp. 359–362.

M. Drieschner (ed.), *Carl Friedrich von Weizsäcker: Major Texts in Philosophy*, SpringerBriefs on Pioneers in Science and Practice 23,
DOI: 10.1007/978-3-319-03671-7_2,

phenomena. The majority of my considerations will be of a phenomenological character in this sense. The problems of phenomenology are different from those of physics. Whereas physics struggles foremost with the complexity of its subjects, it is the simplicity of phenomena that poses the greatest difficulty for the conscious mind reflecting on them, as reflection on givens certainly does not follow along its natural line of thinking.

The second part is set under the heading "Regional Disciplines". This means the individual areas of classical physics as well as neighboring sciences. This section may be considered relatively the most secure.

The third part deals with "Elementary Objects". I have designated as such the subjects of relativity theory and atomic physics. Entirely different objects form the basis of phenomenal givens. Objects are what make a real understanding of phenomena possible but they are only found at a high level of abstraction. How both lines of inquiry relate to each other will be presented at length in the following.[4]

Our science is strongly influenced by the *deductive* disciplines of mathematics. A few statements, called axioms, are presupposed there; all other statements should follow from them. Formerly, axioms were regarded as evident; in most recent times they are often treated as preconditions without any assumptions being made about their truth, consequently turning the whole system into an 'if–then' logical construct.

Physics obviously is not made that way. The concept of *inductive* science comes closer to its essence. Individual statements made by experience are the immediate givens, out of which the few simple principles are gathered by systematic generalization. The finished inductive structure can then perhaps be recast in deductive form at the end.

This image comes closer to the reality of our science but it does not contain decisive features yet. The words deduction and induction both conjure up the simile of a pyramid for science that is either resting on its apex or culminating in an apex. Recall, for comparison, our disposition with the three parts: Elementary Givens, Regional Disciplines, and Elementary Objects. In this simile, science has *two* apexes. Physics does indeed permit a dual structure.

One can start out from the elementary given, from concepts such as number, time, space, thing, cause, motion. This structure finally leads to the atom, like to an outermost twig of a branching tree. This may be called the *phenomenological* structure of physics.

One discovers, though, that concepts like atom, field, or wave function yield a new substantive unity that even questions phenomenological concepts. The true link between phenomena is only revealed when penetrating behind phenomena. A different kind of *substantive* structure of physics is implied.

[4] This text comprises only the first section of the first part of the introduction to the book, cf. footnote 2—*Editor* [MD].

Which structure is the true one? We cannot do without either of the two. The only way to the objects leads via phenomena; understanding the phenomena is only attainable by objects. A *mutual dependence* exists between both ways of construction.

In addition the two apexes are not the most certain part of the system but the most uncertain part; they are like mountain peaks piercing into the clouds. This is clear for the substantive apex. It lies beyond immediate sensory perception. It is merely a thought-up or hoped-for point; in reality, there is only a *frontier of research* on the substantive side; it has even been said that the substantive apex is not completable. The phenomenal apex is in no better a position, though. Its simplest concepts, such as space, thing or causality, impinge on the area of philosophy and, whatever its merits otherwise may be, philosophy is famous as being the science with the most persistent and unsolvable controversies. One would be pleased just to be able to find a generally acknowledged research frontier in it. Only the centre of the double pyramid is barely free of controversy, its belly of science, classical mathematics and physics: Euclidean geometry, arithmetic and analysis, mechanics, thermodynamics, electricity, optics, etc.

Any attempt at knowing the real runs up against similar difficulties. How does one address them? Theoretical philosophy of science has not yet made available any concepts by means of which we could contemplate this situation adequately. Therefore, let us initially fix them by means of *analogies*. Heisenberg says that the closed disciplines of exact science seem to float over unexplored depths on every side. One could also compare science with a ship that "is midway" between the unexplored heights of the sky and the unfathomed depths of the ocean. And if we wanted to add the progress of research into the picture, we could choose a less poetic simile: Science is similar to the task of unravelling a tangled ball of yarn, just a few threads of which are exposed at its centre, whereas we have none of the ends in hand.

The simile of the ball of yarn allows another application: Maybe the two ends are connected together. I spoke about the mutual dependence between the two ways of construction. It is most distinctly manifest at the pertinent frontier of research. The most modern physics of objects imperceptible to the senses has not only stimulated but also used considerations on the foundations of sensory experience. Think of the concepts of simultaneity in the special theory of relativity, of materiality and causality in atomic physics. No matter how the ultimate structure of physics will look, if it should ever come to that, it owes its formation to repeated runs through the *cycle* of mutual dependence between our phenomenal and substantive concepts. To speak metaphorically, the double pyramid repeatedly meets itself end-to-end like a ring.

The foregoing was said in order to raise awareness about specific problems. It itself is not yet part of the conceptual structure, as the utilization of metaphors shows. I myself would like to express again in a metaphor what these metaphors teach us about the process needed for this conceptual structure. It is an anecdote.

Niels Bohr is the man from whom all atomic physicists learned the kind of thinking that I have been trying to indicate by those similes. He comprehends under the term 'philosophers', perhaps not entirely rightly, primarily people who have not grasped this floating character of knowledge and want to construct all knowledge out of one fixed point. One time, we were in a ski lodge together and were washing up the plates and glasses after a meal we had prepared for ourselves. Bohr was particularly lovingly polishing the glasses dry and proudly observed afterwards how clean they had got in his hands. Then he said thoughtfully: "If a philosopher had been told that dirty glasses can be cleaned with dirty water and a dirty towel—he wouldn't believe it."

We must, in fact, start with the impure concepts that practice offers us and spend time cleaning them by rubbing them against each other, so to speak, without even foreseeing the end of this polishing. I am perhaps not entirely in agreement with just one point in Bohr's formulation, that the very awareness of this provisionality, this indicative character of every concept, does not seem to me to be genuine philosophical awareness.

I start by attempting to repeat in epistemologically precise terms the essential content of what I have just said in similes. Henceforth I shall strive for rigorous conceptuality. It lies in the nature of this subject matter that these concepts, too, will bear a character of provisionality and imprecision in order to be immediately understandable. They are introduced to lay the basis for their own mastery.

2.2 Cognition

The view that I would like to question thinks that somewhere in science there could be absolute, intrinsically fixed certainty. Absolute certainty could also be circumscribed by the words: Cognition subject to no doubt. Thus the concepts *cognition* and *doubt* become the object of examination.

Let us consider one simple piece of physical knowledge, e.g., "Lead is heavier than water."

This statement is right. What does this mean?

The statement *asserts* something. That which it asserts is a *matter of fact*, namely, that lead is heavier than water. The statement is right if the matter of fact *holds*, i.e., if lead really is heavier than water. Lead is, in fact, heavier than water, and that is what I initially mean when I say that the statement is right.

The matter of fact would hold even if I had not asserted it. I have now asserted it, however, because I have *recognized* it. I *know* what I have recognized. This cognition or this knowledge is *expressed* by that statement.

The statement therefore refers to two things: to a process or state in my *consciousness*, which I call cognition or knowledge, and *to what* I *have* consciousness of, the matter of fact. Consciousness is *consciousness of something*. I call the individual process of cognition or state of knowledge an act of consciousness. The matter of fact, I call the *content* of the cognition or knowledge. I say that the statement *expresses* the act of consciousness and *asserts* the content of consciousness.

I say that to me the content of each individual act is *given*. I thereby also express at the same time that to me just the content of the act is given, not the act of cognition itself, in any case not explicitly. If I say, "lead is heavier than water", I mean that lead is heavier than water and nothing else. I don't mean that I am thinking and just now know that lead is heavier than water. I am certain, however, as I am posing the question whether I am just thinking it and know it right now, that I am just thinking it and know it. The consciousness is not unfamiliar with itself as a consciousness, but it does not naturally take this theme up. The consciousness knows its content *explicitly*, but knows itself *inexplicitly*. Consciousness is generally *oblivious of itself*. It "thinks of" the content, not of itself.

If I want to explicitly recognize consciousness, I must perform an act of cognition that *asserts* the existence of what was *expressed* in the original statement: the cognized knowledge. I call this new act of cognition an act of *reflection*. By it the consciousness is 'bent back' upon itself. I call this new act of cognition *reflective cognition*. The original act, I call a *plain* cognition. A piece of knowledge or cognition that has become a reflective cognition—in brief: on which I have reflected—I call *reflected knowledge* or *reflected cognition*. If I understood under 'cognition' a new process each time, every cognition would again always be plain. By my cognizing that it 'recognizes' an already familiar matter of fact, however, it can participate in reflected knowledge. Reflective cognition is generally plain unless it is reflected on once more.

As long as I just think of the original content, that lead is heavier than water, I can say the statement "lead is heavier than water" *is* the cognized knowledge. When I reflect, I notice that the statement actually only expresses this cognized knowledge. That is, I now distinguish between the *statement's body* (this sound, these chalk strokes on the blackboard) and the *sense* of the statement. The word 'sense' is ambiguous, as it can mean either the act or the content. I shall therefore only use it where the distinction between act and content, between expression and assertion is of no import.

Distinguishing between the statement's body and its sense is an act of reflection. Usually the statement self-obliviously *serves* as an *expression* of the act or as an *assertion* of its content, which according to our definitions is equivalent. A statement that serves in this way I call a *plain expression*, a *plain assertion* or, in short, a *plain statement*. A statement whose sense has been reflected on, I correspondingly call a *reflected statement*. One can likewise use individual *words* plainly or reflectedly. I do not discuss the details of these possibilities here; they would lead us deep into logic.

An act of cognition does not need to be expressed. I can cognize a matter of fact silently but consciously. I let a piece of lead fall into water and see it sink down; now I establish or remember that lead is heavier than water, but it is not worthwhile to speak about it. The thought can also come up or be evoked on the side, while I am doing something else. I can weigh down a sack with lead so that it sinks. Here my matter of fact is *inexplicitly carried along* within a context. I make use of it without expressly thinking of it. We do in fact constantly take into account an inestimable amount of matters of fact in daily life that we do not give any special notice to. Cognition therefore is not attached to expression.

Where among this series of phenomena is the boundary beyond which one cannot speak of cognition? Such a boundary will not be determinable without some arbitrariness. I do not see in this a weakness in the cognition concept. This fading of cognition, of consciousness in a series of diminishing degrees of expressibility is a phenomenon that we must look directly in the eye.

Each act of expressed consciousness is surrounded by a *corona of unexpressed consciousness* that fades away into the completely unconscious. (For instance, the visual field has a focus of attention, the relevant matter of fact that is fixed upon, around which is a multifarious corona whose conscious character of perception fades away outwards toward the limits of the visual field. The limits of the visual field are blurred and with heightened sensitivity can be expanded astonishingly.) If one denotes as cognition only what is expressed, this corona is not cognition. If one denotes as cognition the *apprehension of matters of fact*, then there is *unexpressed cognition*, indeed I would even volunteer to justify speaking of *unconscious cognition*. In the following I want to choose a manner of expression, according to which any apprehension of matters of fact is cognition, and distinguish explicit cognition by this adjective.

Any explicit cognition presupposes an abundance of inexplicit cognition. If I establish: "This piece of lead sinks in water", I have inexplicitly also been thinking: "this piece of material is lead", "water is inside this pot"; I threw the lead into the water, thereby silently applying the physical facts of free fall, the physiological exertion of energy required for carrying, the conscious fact that I want to perform an experiment and what for, etc. And how do I know that this material is lead? Because it had been given to me as lead, because it is gray, heavy, soft. I know each of these facts and this knowledge has a prehistory. It could be another rarer element. But my informant is not deceiving me. I contemplate: No, he has never deceived me yet. Thus every expressed act is *embedded* within an inestimable multitude of unexpressed acts.

Any inexplicit cognition is plain. One can hence also say: Any expressed cognition presupposes an abundance of plain cognitions, without which it would be impossible.

2.3 Doubt

I don't know everything. There are matters of fact that I do not know. Just because of this, special acts of cognition are necessary. I must *seek* cognition.

This search can *fail*. Either it can fail in a way that I know I did not find the sought item. Then I at least know one matter of fact: that I don't know. Or else it can fail in a way that I don't know that it has failed. Then I am of the *opinion* that I have cognized something, but illegitimately. My act is then thought to be cognized knowledge but it is a *mistake*.

Any act that is meant to be cognition I call *cognition intention* or *intended cognition*. A cognition intention that really is cognition I call *true*. A cognition intention that is a mistake, I call *mistaken*. The statement that expresses a cognition I call *right*. The statement that expresses a mistake I call *false*.

He who errs does not know that he is erring. How should we distinguish between cognized knowledge and mistake? This question sets me before the third option: The intended act of cognition can have the outcome that I don't know whether it has succeeded or failed. It presents to me the option of *doubt*.

True and mistaken cognition intentions can be plain. A cognition intention that is doubted must therefore itself become the object of cognition. It is either eliminated or, if it continues to hold, it is henceforth attended by the cognition intention: "this cognition intention is true" and is, to that extent, reflected.

A statement that asserts a matter of fact, whose tenability is under doubt, I call *doubted*. Different acts can be directed at a doubted statement: doubt, inquiry, supposition, fiction, etc. Doubt first gives occasion to regard the matter of fact asserted by a statement separately from the cognition intention directed at it, that is, to regard a statement as something that can be right or false. Logic is based on this interpretation of a statement. Logic is the doctrine of *dubitables*. (An omniscient being has no need for logic.) Logic, as cognition about cognition, has formed its concepts as suits its nature, out of reflected cognitions. Let us not go into those problems here. Let us just remember that a plain cognition intention, especially if it is inexplicit, is not meant as something that could be true or mistaken, rather that the matter of fact in it is simply given.

At this point it can become clearer why I have defined cognition so that any apprehension of matters of fact be understood as such. One can doubt the tenability of a matter of fact irrespective of how explicitly it has previously been apprehended. If the apprehension was inexplicit, it is raised by the doubt itself to the level of explicitness. It is desirable to draft the concept of cognition broadly enough that it encompasses all dubitables; so that every act of doubting corresponds to an act of cognition intention that it is doubting.

How is doubt eliminated?

Someone may doubt that lead is heavier than water. I take a piece of lead and throw it into the water. It sinks down, therefore lead is heavier than water.

This is convincing but only to whoever accepts an abundance of plain cognition intentions as cognized knowledge. He must believe what he sees. He must be certain that this is lead, that that is water. A prestidigitator could deceive him. He could be dreaming. Hence, the elimination of doubt is, like any act of reflection, linked with plain cognition.

Doubt that reveals some obscurity of expression is a special case. A small piece of lead is lighter than a large pot of water. One has to say more exactly what 'heavier' means in the statement: i.e., 'specifically heavier'. A *definition* is given. That is, reflective manufacturing of an expression indicating the sense that a word frame should hold. This sense must be denotable by other words, however. It therefore already presupposes plain expressions, e.g., here 'volume', 'equal', etc.

All doubts discussed up to now referred to a *single* cognition intention or, if at all, to an individual range of cognition intentions. It also arises generally out of a particular individual *doubt motive*. We express this referencing to particular cognition intentions and doubt motives by calling it a *relative doubt*. Its elimination leads to a *relative certainty*. It is, on one hand, certainty about an individual matter of fact or range of matters of facts, on the other hand, certainty only about the particular individual doubt motive that has been eliminated. The reflective cognition eliminating the doubt cannot yield more certainty than inherent in the plain cognitions out of which it is composed or which it makes use of.

The point of departure of our entire consideration was that we wanted to acquire concepts suitable for judging a particular ideal of cognition: that of *absolute certainty*. The concept of science that became questionable to us in the introductory considerations wanted to see at least some cognitions secured against *any possible* doubt. Is there such a thing? Do cognition intentions exist whose truth cannot be doubted at all anymore? Do statements exist whose meanings are indubitable?

We are far from being guided to absolute certainty by this question. On the contrary, it opens for us the possibility of *absolute doubt*.

He who errs does not know that he is erring. This single statement basically lays out the impossibility of absolute certainty. Yet we have now delved too deeply into the problem of certainty to be able to be satisfied with a single statement as an answer. We ask whether absolute certainty doesn't exist somewhere nonetheless? This question is our guiding thread for learning the *art of doubting*. We must understand something about it in order not to be crushed unsuspectingly by the *fate of doubt*.

One can speak of an art of doubting because a person living normally along is not versed in doubting. His understanding of the world is based on a plain apprehension of matters of fact, plain cognized knowledge in the broad sense of the word 'cognition'. When he does doubt something, he is prompted into doing so by a doubt motive that itself is a plain cognition. Mr. Mayor is walking up the street toward me. But he's wearing a brown hat. Mr. Mayor doesn't wear such hats. Maybe it's not Mr. Mayor at all. Here plain apprehension of a matter of fact, namely, the hat's brown colour, is the doubt motive. If I doubted everything from the start, I would also have to doubt whether the hat is brown; the doubting motive itself would not be plainly given. In normal life doubt is an individual event that is only possible, the way it is played out, because of the plain givenness of the not-doubted.

One can call the manner of givenness of the not-doubted *plain evidence*. Everybody knows that plain evidence offers no absolute certainty. Appearance is misleading and the difficulty is that one does not know where semblance is involved. But in practice one mostly manages to attain the certainty needed in life, which if preceded by doubt can be called *reflected evidence*. Philosophers sometimes think that reflected evidence could be elevated up to *absolute evidence*. But how do they defend themselves against the statement that he who errs does not know that he is erring? Classical examples prove that the evidence experience can be misleading.

It is sensorially evident that the Sun orbits around the Earth. But the opposite is true. It is evident to pure intuition that parallel lines don't intersect. But doesn't non-Euclidean geometry enter the discussion? I recently dreamed that two times two equals five. Now that I'm awake I know that two times two equals four. But then I knew the contrary. On what does the certainty that I am now right base itself? A *dream* is the big example of the dubitableness of evidence having dawned on people. What if life as a whole were a dream? What if a god were systematically deceiving us?

Is this question already absolute doubt? Descartes tried to develop absolute certainty out of it. I doubt everything. I doubt. I. One thing is certain, that I doubt. Doubting is a way of thinking. One cannot think if one doesn't exist at all. Therefore, I cannot doubt my own existence. Cogito ergo sum—I think, therefore I am.

This train of thought is extraordinarily important because it directs the view onto what has been called pure consciousness. It is a first attempt at the undertaking, pursued up to the so-called phenomenological reduction by Husserl in our century, to distinguish conceptually sharply between the consciousness and its objects. Here, though, we are just concerned with how it relates to doubt. And there the statement to be made is: It offers no absolute certainty because it does not presuppose absolute doubt. Descartes was still a beginner in the art of doubting.

Descartes doubts, for instance, the *rightness* of statements but not the *meaning* of his statements. He asks: "Does the world exist?" and dares to doubt whether the answer must be 'yes'. He doesn't ask: "What does the word 'exist' mean?" "Does it mean anything at all in connection with the word 'the world'?" Is 'to exist' perhaps a concept that only makes sense relative to a particular context? When *Romeo and Juliet* is performed, Juliet exists as certainly as does Romeo. But they exist only 'for the play'; 'in reality', Juliet is Miss Miller. To Homer, Zeus exists and works wonders; to the modern physicist, the atom exists and works wonders. Is the atom less of a myth than Zeus? Is there a different existence than 'existing to' someone? Then the purported certainty by Descartes might be nonsense, because it would be an answer to a senseless question.

Perhaps I don't mean all that I have just said so seriously? How do you know that? You must in any event make it clear to yourself once that one can question like this.

Descartes does not doubt the sense of his statements. He thereby acts, on a higher level, exactly like the person living unreflectedly along. Certain elements of cognized knowledge are plainly given for him, so it does not even occur to him to doubt them; and specifically these become the motive for him to doubt other elements of cognized knowledge. Just because he believes that the statement 'the world exists' makes sense does he believe one could doubt whether the above statement were right. When a statement makes no sense, one doesn't ask whether it is right or false. But far be it from me to positively assert that the statement 'the world exists' made no sense. That would be badly doubted. I just say that one could doubt whether it makes sense.

All that I have been saying does not express absolute doubt. One cannot express absolute doubt. Whoever utters it does not presuppose the plain sense of the words. Absolute doubt can only remain silent. It cannot be argued against, precisely for that reason. It cannot be argued in favour of, for that reason either.

One could say, though, that absolute doubt is unattainable for a living person. This is right. Apprehending matters of fact isn't just a theoretical process. It is constantly necessary in our lives; and, no matter which way we turn intellectually, animal life puts it ever at our disposal. The doubting philosopher who is startled back into reality by a wasp sting or a box on the ears is a favourite theme in a comedy. A statement that is almost the converse of the Cartesian statement states: *He who lives doesn't doubt all.* It is made to take on the Cartesian form by specifically saying: To be able to doubt, one mustn't doubt all.

You would be completely misunderstanding me if you thought I wanted to hold my ground against absolute doubt, nevertheless, by these last considerations. He who still lives doesn't doubt absolutely yet; but who says that the living are right? The art of doubting obviously ends here. One cannot propose to doubt absolutely. But one can be driven onto a path that offers no foothold against absolute doubt. There is no argument against the fate of doubting.

This doubting isn't an intellectual enterprise anymore.[5] It is a form of despair. With Kierkegaard, in Dostoyevsky's *Ivan Karamazov*, in Hoffmannsthal's *The Lord Chandos Letter*, you will find more about that than with all the philosophers. Its utmost point will never be written down, though. Wanting to be right against it is not only impossible, it is also wrong. The despairing soul will never be reached by being right anymore, perhaps only by love.

2.4 Faith

The discussion about doubt is framed by the two statements: He who errs doesn't know that he is erring, and: He who lives doesn't doubt all. For us who are alive there is neither absolute certainty nor absolute doubt. It surely cannot be denied *that* we find ourselves in this situation. Yet we find ourselves in it even with a relatively good conscience. We have considerable trust in what we know, and don't think we are doing badly by it, despite standing alongside the abyss of possible doubt. We must try to find concepts that describe this attitude distinctly.

[5] Note 1983: When I quoted this statement: "He who errs doesn't know that he is erring," to my uncle Viktor von Weizsäcker in 1948, he spontaneously rebutted: "That's not right. He does know it, all right, but he doesn't want to know it." Thus a discussion was opened that did not feature in this lecture; one can say: the moral aspect of epistemology. The closing passage of the section on doubt is rather commented on by it. Despair is not just despair about my knowledge and known reality but also about my will and the willed good. Ivan Karamazov did not despair about God's existence but about His kindness; Dostoyevsky saw this as the deeper atheism.

I would like to choose the word *faith* for the attitude we have toward the contents of our knowledge, in view of the two impossibilities of absolute certainty and absolute doubt. We must agree on the exact meaning in which this word should be used.

In general, faith is considered to be deeming something true that one does not know. Faith and knowledge are then regarded as opposites and even relegated into separate fields: religion and science. I consider this whole confrontation false and chose the terminology that I shall explain now primarily in order to make slippage into this interpretation through word usage impossible.

Faith is not an intellectual act but a way of living. Having faith in something means behaving in every situation the way one must behave as if what one believes really existed. Holding something to be true is merely the intellectual pinnacle of believing conduct accessible to reflection. To express it in a metaphor: A soccer player must occasionally pass the ball on, to another player of his team. This only makes sense if he can count on his partner taking over the ball and, perhaps, kicking it back to him. There is no certainty about that, since the other player could be impeded by the opponents or may miss the ball. Despite that, he must pass it on to him. This calculated passing of the ball on to his counterpart and expecting it back despite the uncertainty is faith.

Faith—just like cognizing—is conduct toward a matter of fact. If cognizing is addressing the matter of fact as a given, then faith is addressing the matter of fact independent of whether it actually is given. The fact that neither absolute certainty nor absolute doubt is possible for us can also be expressed this way: *One cannot cognize without faith.* This may become clearer if we distinguish between *inexplicit and explicit* faith, as we did with cognizing.

The deeper we go down into the sphere of inexplicitness the more impossible it becomes to distinguish at all between, on one hand, knowledge and cognition and, on the other hand, faith. In order to establish that lead is heavier than water, I let a piece of lead fall into water. By letting go of it, I count on it dropping. I cannot know that with certainty in advance. If it were paper, it could be drawn sideways by a gust of wind; if it were iron, by a magnet; maybe it had been lying in tar and will stay stuck to my hand. Maybe a new effect still unknown to physicists will occur. But I cannot be detained by such scruples. I let go of it and count on it dropping; and almost always what ensues will prove me right. This "counting on" is the passing of the ball and, to that extent, faith. It occurs with such a great chance of success, however, that it does not need any concentration of attention. Just because it is almost a piece of knowledge, it can remain inexplicit. The statement: "Within the sphere of inexplicitness, knowledge and faith are not clearly distinguishable," can hence also be read conversely: "Where it does not become necessary to distinguish between knowledge and faith, conduct on a matter of fact can remain inexplicit."

If I reflect on whether I know a matter of fact or 'just' believe it, I have entered into the sphere of explicitness. I encounter a mistake. That is, I see that much of what I had unexpressedly believed, I did not in truth know—that it was false. The striving after absolute certainty was an attempt to make faith superfluous.

This has proven to be impossible. By living, we believe. Because we know this, it now involves an expressed faith. *How* do we believe, or *what* do we believe in, now?

It would again be a misunderstanding stemming from reflection to want to try to formulate a 'legitimate content of faith'. If one could prove the 'legitimacy' of some content of faith, one would probably better be speaking of knowledge. We just have the matter of fact as the point of departure, which is summarized in the statement: "He who lives, believes." We do not ask what he *may* or *should* be believing, rather what or how he *in fact* believes. But this is different for different people.

The person who does not give much thought to these questions I will call the *natural person*. He encounters relative doubt from time to time and is satisfied with reducing it to relative certainty. He notices that one does not get far with fundamental doubting and lets such things be. His faith is *inexplicit granting* of the plainly given.

Also someone who has deeply engaged in doubting and perhaps even experienced despair finally finds himself, if he lives on, facing the same world again that was a given for him as a natural person. He will then have learned caution on many points and relative doubt; maybe the floating character of all knowledge has become clear to him, the possibility to doubt all. But by living he lets the world be. This is *explicit plain granting*. One can hardly say it more clearly than Faust at the instant he returns from despair: "The teardrop wells up, Earth has me back." The tear is the real that he plainly grants, and along with it the world, since weeping means living.

Whoever has come back out of real despair probably always has undergone an experience in the range called religious. The possibility of living on is mostly attached to this experience for the person. His continued life thus becomes a conduct that reckons with a reality shown by this experience in the manner of faith, even if this reality is not or is no longer manifest. *Religious faith*, where it is genuine, is therefore in a special way not merely a taking-as-true but a way of life. It is not a mere granting of something anyway manifest but an active, constant address or appeal to something not manifest outright.

I have attempted to *describe* some manners of faith. I did not try to argue about its value, since one can only do so by oneself believing, hence not from a place lying beyond the decision in any conscious faith. This abstention of mine cannot be more misunderstood than by being taken for an expression of relativism with regard to the truth of the relevant substance of faith. I had defined faith in such a way that cognition is not possible without faith. Faith is, consequently, the way to truth; and precisely because it is the only way to truth, one must engage in faith if one wants to evaluate truth. This is valid at the simplest levels: Whoever does not grant the verdict of the senses, with him one cannot speak about matter. It applies just as well in religion itself; Christ says: Whosoever does God's will shall know whether my doctrine be of God (John 7, 17). If in search of the whole truth, it is therefore impossible to make a philosophy independent of the religious decision.

A philosophy that purports to be independent of faith is, in truth, just not conscious of the faith proper to it.

The subject of this lecture course, physics, does not compel us in any obviously ascertainable way to reach a decision on the ultimate questions of faith. For, the faith that physics has as a precondition, faith in the applicability of rational thinking to sensory experience, is a common property of people of our times. (I only need to point to technology. It is perhaps its most conspicuous manifestation.) One might almost even say, the faith of physicists is the sole faith connecting all people of our day.[6] Thus we do not need to produce this faith but instead can begin right away to examine the contents it gives us. It is different when we ask what the significance is that this faith has become possible and generally prevalent. As living human beings, we cannot escape this question as well. However, I do not pose it at the beginning of this lecture but at the end.

2.5 Methodological Consequences

Now we shall pin down the *principles* of our *method*. There is no absolute certainty as a point of departure. We must presuppose a faith. We want to speak of physics. Therefore we presuppose the faith of physicists: What is this faith composed of and what does 'presuppose it' mean?

I understand the confidence in the methods and results of physics necessary to conduct physics as physicists' faith. I repeat that faith is not (or not just) a taking-as-true but a way of living. Presupposing the physicists' faith hence means, put in human terms, granting validity to physicists. One might not grant them validity in what they do and think outside of physics. But presupposing their belief means granting that they are doing things about right in their own field. What they have to say must be taken seriously because otherwise one cannot converse with them at all.

I have purposefully expressed myself very vaguely now. But we want to come far enough along to formulate opinions; thus we must strive for conceptual rigor. For that it is necessary that we define the concept 'presuppose' more precisely. I could also describe this act as *reflected granting*. What does that mean?

By considering the methods and certainty of physics we have entered into a field that one would not call physics anymore but philosophy. By having once gone down the path of doubting, we have left the plain faith of the natural person in his surroundings, the plain faith of the physicist in object and method. By then recognizing that no knowledge is possible without faith, we coined the concept of a physicists' faith. We *reflected* on this faith. This reflection is somewhat different from plain—even if explicit—granting of faith, upon which our life depends. This latter faith makes it possible for us to analyse something but it does not itself

[6] I later took this notion as the point of departure of the lecture *The Relevance of Science* (London: Collins, 1964).

become the object of analysis; we have already seen that we could not come to any agreement otherwise, because, assembled together as we are, we do not have the same faith in many decisive things. Notwithstanding where each of us gets the energy to live, we want to grant the physicists' faith as something that exists in order to examine it.

Permit me to present a metaphor, albeit saying at the outset that it does exaggerate in one respect. The frogs trapped under the winter ice of a pond promised to sing like nightingales if they were freed. When spring arrived, they sat along the edge and croaked as in the old days.[7] Croaking is the frogs' plain faith. They can just either live and croak or not live. We, however, don't want to croak as in the old days. We only want to grant that frogs exist and that they croak and want to observe how far croaking can take one.

Physicists are the frogs, and our decision to grant the physicists' faith is the granting of croaking. But this metaphor exaggerates the distance between the grantor and what he grants. The physicists' faith is part of the faith of all people of our times. Whoever switches on an electric lamp concedes in this way the inexplicit faith that he expects physics to be right about its evaluation of practical tasks in life. To that extent we analyse our own faith, which we cannot give up at all, within the physicists' faith. We all are 'frogs.' On the other hand, we aren't going to be prepared to follow this faith without reservations as an absolute truth. We reserve for ourselves the possibility of doubt in each individual case.

This doubt can only be meant as relative doubt. Otherwise, as an absolute doubt, it would be retracting the matter we want to have granted. The boundary between relative and absolute doubt itself cannot be drawn absolutely, though. No barrier can be indicated beyond which relative doubt ought not to be extended. We have no other alternative to participating in the intellectual movement in physics and experiencing along with it how each relative doubt might not ultimately overturn concepts but rather clarify it further. Nothing else was meant by Bohr's metaphor of cleaning glasses.

I occasionally call this reflected granting a *hypothesis* of physics. We presume that physics is cognized knowledge and observe what comes out of it. Now, what does the practical procedure look like that must be followed?

What we are granting is not a small, sharply outlined area of doctrines, not the apex of a pyramid but a *way* of *cognizing* and the *breadth* of the *pool of observations* that this way of cognizing conveys. This pool of observations is not sharply defined and not fully organized conceptually. It permits two directions of further inquiry that I would like to call *substantive* and *reflexive*. The substantive inquiry seeks to *expand* the pool of knowledge, the reflexive one seeks to analyse it, to

[7] Goethe: "Ein großer Teich war zugefroren ..." ["A large pond was frozen over ..."]. Johann Wolfgang von Goethe: *Poetische Werke* [Berliner Ausgabe. vols. 1–16], vol. 1 (Berlin 1960 ff): 603.

clarify it. Science moves forward only by an interplay between both ways of inquiry. Here, however, we must regard them as separate methods.

Substantive inquiry can plainly, indeed, inexplicitly presuppose the physicists' faith and asks further in that sense: "This I already know about nature. What else can I find out?" Thus it requires no special methodological contemplation. The reflexive inquiry, by contrast, explicitly performs the *hypothesis*. It asks: "If one grants physics as cognized knowledge, what has thereby already been conceded? What presuppositions are contained in physics?" Hence, not the objects of physical knowledge are made the subject of new knowledge, but physical cognition itself. It is reflection.

Reflexive inquiry is almost what Kant calls the transcendental question: "How is physics possible at all?" That which one already concedes by granting physics as cognized knowledge is the *a priori* of physics, the condition for the possibility of physical experience. I avoid the Kantian expressions, however, because they are stamped in the mentality of absolute certainty which we cannot presuppose. We shall be comparing our notions against Kant's throughout this lecture.

You notice the relation between the two paths of inquiry and this lecture's organization. Regional disciplines are the core of the stock of physics that we grant in the sense of hypothesis. The plain faith of a physicist guides him further along the substantive path of inquiry up to what he regards as the most elementary objects known until then. Reflection, conversely, leads him to delve into what, according to his own knowledge, must count as the most elementary given. These two motions are the arrows that signify the two "apexes" of physics, which we spoke of at the outset; they constitute the two frontiers of research.

Another confirmation is needed that reflection also creates a research frontier. One could say: "A given simply is given. One only has to notice it and then it is known." However, the self-forgetfulness of cognition is overlooked. The eye sees things but not itself. The consciousness is explicitly consciousness about some content and only in an inexplicit way hardly known to itself, consciousness about itself. Reflection simply means establishing what is actually given for us and how it is given for us, therefore, making the 'phenomenon' in the cognized knowledge conscious. To that extent, the basis of the reflexive method should be denoted as *phenomenology*.

Phenomenology is at least as difficult and unfinishable as physics because reflection is a line of inquiry unnatural to the original consciousness. If phenomenology were an end in itself for us here, we would have to develop it by increasing reflection, setting out from the regional disciplines. But we want to establish physics here.[8] Phenomenology is merely an auxiliary science for us. If we inquire about the given in normal physical knowledge, the aim is not to examine cognition but the given inherent in it. That is why I place the phenomenological chapter at the beginning.

[8] i.e. in the lecture this text is a part of—*Editor* [MD].

This doesn't mean that I could start with the simplest givens independent of all that follows. It just means that I throw myself into the swirling current at one place and then let it carry me around. In order to make understandable the first givens I am going to be speaking about, I shall employ a language that relies on your already having the physicists' faith and already having reflected on it to a certain degree; otherwise this language would remain incomprehensible. This is expressed in that vocabulary is used in the description of the first phenomena to denote phenomena described only later on. Herein is expressed the *cycle* in which any cognized knowledge is acquired.

We must now try to define this cycle conceptually more precisely. Any matter of fact that is the content of a piece of knowledge that I do indeed have, I call *given*. Any matter of fact that holds, whether or not it is given for me, I call *factual*. I am aware that these labels are also liable to raise many doubts; but at the moment they should serve as a brief form of expression for initially pointing out a phenomenon. From the standpoint of strict methodological doubt, I may only assert a given. However, faith always presupposes the factual, which is not-given, for me: He who lives, believes; hence presupposing the not-given factual is a precondition for living. The substantive line of inquiry keeps within this faith and tries to understand the given in the totality of the factual as well. On the other hand, I can only really know about the factual insofar as it becomes given for me; relative doubt is always permissible, of course. Hence, methodologically, the factual should only be characterized as what *can* become given for me. With this 'can' enters the concept of *potentiality*, which will soon become a main object of reflection for us. Thus we have a cyclical connection: The given is an excerpt of the factual, the factual is what can become given. Otherwise put: The individual is only understandable from the whole, the whole is only exhaustible by way of the individual. Therefore the methodological necessity to jump into the swirling current.

I have, in fact, already done so in this methodological preliminary consideration. I relied on your having often cognized knowledge, having had doubts and faith already, and know what it means; and then I tried to maintain the degree of precision on the problem I had started out with in going around this cycle. That is also why I first went through a purposefully vague cycle with the summary overview and the metaphors. The concepts used there gained more precise meaning by the methodical contemplation we are just completing: the second round. The first round was only possible because the matters of fact exist that were drawn into closer view in the second round. The given in the second round was the not-given factual in the first round, but which was the factual condition for the given in the first round. Now we are entering a third round which will comprise all the rest of this lecture course. It will acquaint us again with the factual conditions of the given in the second round and thereby make the second round understandable, in a certain sense. We would have had difficulty understanding the process we shall be undergoing in the third round and hence would have hardly been able to handle it correctly without having gone through the second round beforehand.

There is no need to comment that these circles do not demonstrate the path by which these insights were first made but only the shortest form that occurred to me to present the acquisition of knowledge. All knowledge is gained by struggling through by trial and error.

But enough of this methodological business. Let us turn to the subject matters.

Reference

von Weizsäcker, Carl Friedrich, 2004: *Der begriffliche Aufbau der theoretischen Physik, edited by Holger Lyre* (Stuttgart–Leipzig: Hirzel).

Carl Friedrich von Weizsäcker at the age of 85, painting by Lili Bansa in 1997. © Lili Bansa who granted permission to use this photo

Chapter 3
A Description of Physics

This is an excerpt from the lectures on physics given in a course, "Critique of the Sciences", at the University of Hamburg during the winter semester 1968–1969.[1] From the original transcription I have prepared and further edited the present excerpt, without abandoning the style of oral delivery.

3.1 The Role of Physics

In the framework of the sciences, physics occupies something like a key position. This is true methodologically, systematically, and socially.

Methodologically, physics has become a paradigm for all of science. The conception of science in other disciplines takes its bearings from the conception of science in physics—even when a discipline stresses the difference between its own conception of science and that of physics. Thus, the formulation of a scientific conception of the humanities, attempted, especially in Germany, around the turn of the century, involved defining the essence of the humanities by contrasting it with the essence of natural science. The weakness of this attempt at a definition lies, it seems to me, in its defensiveness, its acceptance of the 'opponent's' terms. Physics served as the paradigm even when the paradigm's sole function was to say why it should not be imitated.

Secondly, physics occupies a central position in the system of the natural sciences—now including the science of life, and the science of man insofar as man is an organism, i.e., certainly medicine. Physics is a central, a basic science. According to the philosophical hypothesis called 'physicalism', all natural events can be reduced to the laws of physics and may thus be viewed and understood as purely physical events; and there is no other science that could make such a claim.

[1] This text was first published in: *The Unity of Nature* (New York: Farrar Straus Giroux, 1980), pp. 84–100; it was translated by Francis J. Zucker from: *Die Einheit der Natur* (Munich: Hanser, 1971): I.6.

M. Drieschner (ed.), *Carl Friedrich von Weizsäcker: Major Texts in Philosophy*, SpringerBriefs on Pioneers in Science and Practice 23,
DOI: 10.1007/978-3-319-03671-7_3,

Whether this claim can be justified will be discussed later in this book (in Part III).[2] In any case, it is a claim frequently made, and even if not explicitly stated it is usually being presupposed.

Thirdly, physics enjoys a high social prestige—if, for example, prestige is measured in terms of the amount of funding a science receives from the government. Nowadays physics is one of the best-paid sciences. Perhaps you will permit me an expression that is not to be taken altogether seriously; it refers to the chicken coop. We know from the animal psychologists that chickens maintain a 'pecking order'. Every chicken knows which chickens it is allowed to peck, and by which it can in turn be pecked; and the chicken allowed to peck all others is the first to get to the food. Perhaps one is justified in saying that physics or, more generally, natural science, heads the current pecking order of the sciences.

A science in such a privileged position has many reasons for examining and criticizing itself; after all, the question of whether this pecking order is justified remains to be answered. I will now discuss the internal structure of physics in order to discover, if we can, the reason for its great success.

3.2 An Anecdote as an Eye Catcher

In 1935 an American experimental physicist, a teacher at one of the best universities, published a paper in which—so he claimed—he showed empirically, by means of a difficult, carefully performed experiment, that in the case of very high-energy atomic processes the Law of the Conservation of Energy does not hold strictly, but only over a statistical average. This paper, read all over the world, was much discussed among physicists, and came also to the attention of my teacher Heisenberg, whose assistant I was at that time. Heisenberg read the paper and said: "He measured wrong!" And indeed, a year later the author was forced to revoke his claim. He published another paper in which he said that an error had unfortunately been made in the interpretation of his data, and that the energy law remains strictly valid in the case he investigated.

I told this same anecdote at our last meeting, in order to illustrate the sureness with which physicists can judge facts of nature, and I then proceeded to describe the effects of this sureness in the technical transformation of the world. Today I want to reverse the question by asking: What happened in this case? How must a science in which such a thing is possible be constructed? With what process are we dealing here? To lend sharpness to the question: we all learn, rightly and appropriately, that natural science, especially physics, is an empirical science; that one has to bow to experience; that experience, while perhaps limited in its capacity to confirm theories, can certainly refute them. And here we had a theoretical law,

---

[2] Carl Friedrich von Weizsäcker: "The Meaning of Cybernetics", in: *The Unity of Nature* (New York: Farrar Straus Giroux, 1980), pp. 227–366; it was translated by Francis J. Zucker from: *Die Einheit der Natur* (Munich: Hanser, 1971), part III.

namely the First Law of Thermodynamics, the Law of the Conservation of Energy, being refuted by experience based on sound scientific method; and the good physicists—Heisenberg was not alone in this—reacted by simply denying the experience and saying: “That is no experience.” They put it this way: “He measured wrong!” And they were right.

Experience itself confirmed that Heisenberg was right. Not only did the author of the paper admit to having, as he put it, misinterpreted his experiments, but similar experiments have been conducted frequently since then and the result has always been that the energy law holds even at these high energies. Heisenberg’s assertion was therefore not based on a lighthearted disdain for experience. On the contrary. It was an assertion which had experience in mind, which dared prophesy that this particular experience would turn out not to be an experience. It is thus unclear what ‘experience’ really means, and when a supposed experience is actually an experience. Physics is based on this. It is based on empirical knowledge which can again and again be subjected to the criticism that it is not empirical knowledge after all. Whoever fails to recognize this structure of our science does not understand what experience and empirical science really mean. But the explication of this structure is by no means simple, and I suspect that there is hardly a single scientist to whom this structure is so transparent that he could, on request, correctly state its relevant and essential features.

It is easier to do science, that is, than to understand it. It is easier to be a physicist and to gain correct physical knowledge than to state what one is doing when one does physics. And the same is true in all the other sciences. I myself have repeatedly experienced this in the following form: As a trained physicist with philosophical interests, I have of course always been curious about what the philosophers said concerning the problems of physics. One sort of philosophical statement, though carefully worded in accordance with the recognized criteria of the human sciences, is nevertheless completely useless. These are statements made by philosophers who, not having studied physics themselves, quote texts written by physicists and comment on them. It turns out that physicists differ from one another, that these texts are frequently quite meagre in their philosophy, contradict each other, and are even self-contradictory. If one then corners a physicist and asks what he actually had in mind in this or that text, he says: “Well, I certainly didn’t mean it the way you interpret it. I meant…” And then one says: “But why didn’t you say so?” He answers: “We physicists understand each other, after all.”[3]

[3] At this point I compared the lecturing techniques of natural and human scientists. (this passage reads: “In a typical lecture in the humanities, the professor reads from a carefully organized manuscript, every phrase of which has been weighed and secured against all sorts of misunderstandings, and the only freedom he permits himself is modulating his voice. A typical lecture by a natural scientist consists of his referring directly to some experiment or to something he writes on the blackboard. He talks in a more or less lackadaisical manner, points at the board once in a while, says ‘this symbol’ or ‘that formula’ and ‘then we find that such and such’, and if one records it all on tape and then listens to it, one understands nothing at all.—Of course I am exaggerating a bit in both cases; one could cite counter-examples for each. But I would say that

It is therefore useless to try to understand physicists by quoting their words; one must know their deeds. Science is based on a know-how that is practiced, and constant reflection on this know-how is unnecessary (perhaps even impossible); the effort to achieve this reflection I would call philosophy.

Unfortunately I cannot systematically explicate the example I have just mentioned, because this would presuppose a greater knowledge of physics than I have a right to expect from most of you. Let me therefore pick an example from the seventeenth century that is more transparent to us; I claim that the methodological problems are the same in our time.

3.3 A Historical Example

Let us recall how Galileo introduced the laws of free fall and inertia.[4] He first derived the law of free fall theoretically and then confirmed it empirically to a rather poor approximation by means of balls rolling down an inclined plane. In the derivation he had to make use of the law of inertia, although neither he nor anyone else had ever seen true inertial motion; for there is no body that is acted on by no force whatever. How such a body would move one can therefore, strictly speaking, not see empirically. One can conceive of it only as a limiting, an idealized case. Galileo had to say, in connection with the law of free fall that all bodies fall with the same speed, which, empirically, they do not. But they would in a vacuum, he said. In Galileo's time one could not produce a vacuum empirically, however. And even today there is no such thing as a vacuum in the strict sense in which Aristotle intended this term—as pure emptiness, which Aristotle claimed could not exist. But the concept of the vacuum helped in expressing the laws of motion in so simple a form that it became possible to analyse, master mathematically, predict, and finally master technically the most complex processes. Modern science as represented by Galileo masters experience precisely by not simply sticking to what experience presents directly.

Galileo was aware of all this. He knew that, when he had the right theory, he saw the phenomena with different eyes than before, and could even produce phenomena which no one would have thought of producing before and which therefore had in fact never been produced. At the same time, Galileo was an engineer. He built not only telescopes, but also a number of mechanical contrivances, including some for military use. Abstractly speaking, this interest in technology is an interest in the production of phenomena that would not otherwise exist. It is an interest in the production of experience and not merely in its passive

(Footnote 3 continued)
the presentations of natural scientists are characterized by a certain nonchalance and by the influx of non-linguistic elements."—*editor* [MD]).

[4] In the original lecture I gave a more detailed account covered in: *The Relevance of Science—Creation and Cosmogony* (New York: Harper & Row, 1964): 103–106.

acceptance. This is no longer the basic attitude of Greek science. To be sure, the Greeks, too, had a technology characterized by marvellous precision. But the idea that the purpose of science is to make technology possible did not hold sway. The dominant attitude was that science contemplates the glory of things as they are, not that science exists for the purpose of changing them. The idea that things will improve if one changes them is not at all obvious; understood in its current, radical sense, this idea is a modern opinion.

3.4 Philosophy of Physics

How is it that the mathematization of natural science, the theoretical prediction of experiments, and the technical transformation of the world are at all possible? So far we have merely posed, not solved, this problem. What must be the nature of things if such enterprises do in fact succeed? After all, it could have turned out that observations stimulate all sorts of ideas, but that none of these ideas enables us to transform the world. Intellectual systems based upon delusion do exist, but most of them betray their delusory character by coming to grief. Delusions are dangerous if they contain a grain of truth and therefore do not come to grief immediately. Could it be that physics is such a delusion? But then the question of how a delusion containing a grain of truth can possibly exist remains unanswered. Or is physics perhaps a real truth? What would we then mean by truth? Let us review some of the answers given to these questions in the course of time.

3.4.1 Theological Platonism

If we consider Galileo—or his great contemporary Kepler, or Copernicus, etc., of all of whom the same is true—we must say that his theory of natural science was a largely Platonic theory. Of the two great philosophical authorities of antiquity known in medieval times, not Aristotle but Plato was invoked, i.e., Plato the mathematician, the follower of Pythagoras.

The reception of Aristotle at the height of the Middle Ages was, historically speaking, the acceptance of an authority which legitimized the yearning for the empirical, for the reality of the secular world. Four hundred years later, in the seventeenth century, Aristotle had become the dogmatized preceptor of bygone times, and men turned, as they put it, away from written books, toward the book of nature itself. Since authorities remained important, it was Plato who, ignored at the height of the Middle Ages, now became the star witness. The Plato on whom the early Middle Ages, as well as late antiquity, had leaned, had above all been the mystical Plato. This was the Plato of the Neoplatonists, the Plato who described the soul's ascent to the divine One. This is a part, but only a part, of the real Plato. Plato also describes and demands the descent back into the world of the senses, back into the cave, as he says in his parable. And with the help of mathematics, in the theoretical part of his thought Plato actually accomplished this descent. This is indicated in the seventh book of *The Republic,* and somewhat more broadly in the *Timaeus;* it must also have formed an essential part of his unwritten teachings.[5] Plato had the design of a mathematical natural science. It was sciences which, from the highest point of the divine, to which the soul is to ascend, sought to explain and represent as far as is possible, the plenitude of all that can be perceived, and which did this by means of mathematical conceptions. This was the Plato invoked by Kepler and Galileo.

Again, this is, at least to begin with, merely an assertion—the assertion that one can understand the world of the senses by means of mathematics. But how can this be? Is it not quite incomprehensible? After all, mathematics deals with ideas. For example, the mathematics of triangles certainly does not deal with a perceived triangle. In earlier lectures on Plato[6] I sometimes drew a circle or triangle freehand on the board and asked: "Is this really a circle? Is the sum of the angles in this triangle really 180°?" The answer, of course, is: 'No.' Neither a triangle drawn freehand nor one drawn with a ruler is accurate; indeed none is accurate that can be perceived by the senses or produced by whatever means. Therefore, mathematics does not deal with objects of the senses. It deals with what we are capable of thinking, what the spirit or mind—in Greek: the *nous*—can grasp. How, then, can our thinking furnish the laws for governing what we perceive with the senses? That is the problem.

Kepler offers a well-thought-out answer. He says: God created the world in accordance with His creative thoughts. These thoughts are mathematical. God created man in His image. Man is an image of God in those attributes that can at all be conceived of as predicates of God. The highest of these predicates is spirituality, spiritual being. As a spiritual being, man is an image of God and can as such rethink God's creative thoughts and recognize them in the material creation. This recognition, this divine service, is natural science.

This is Platonism in Christian language. In purely Platonic terms one would say: The human soul is part of the world soul, and human reason, insofar as it recognizes an eternal truth, partakes of divine reason, which recognizes truth eternally. The world soul, which moves itself and everything else, produces in its movement all that has a beginning and that passes away; and the world soul knows what it is doing.

Kepler is not merely formulating an assertion of philosophical theology; he is also describing his way of doing natural science. The enthusiasm with which his great discoveries filled him made him think of the process of discovery as a viewing of divine secrets, indeed as a divine service. In the history of science, this remained Kepler's personal opinion; unlike his description of the planetary orbits as ellipses around the sun, this opinion could not be taught. But scientists to this

[5] Cf. Gaiser (1963).

[6] Cf. von Weizsäcker (1964: 54–76).

day, especially the great creative scientists, often feel the same way Kepler did. Einstein said: "What else are we trying to do than retrace His lines?"[7] With the progressive decay of religious faith in modern times, the confidence of scientists in the words that Kepler could still use began to wane. It became more difficult for them to express their experience of scientific discovery in such words.

The philosophical problem left unsolved, even from their own point of view, by Kepler and, it seems to me, in the last analysis also by Plato,[8] now emerges more distinctly. It is the actual mediation of the gap between the highest point of philosophy, the One or the Idea itself, and the multiplicity of sensory experience. It is of the greatest significance that the successful discovery of scientific law is *experienced* as a part of this mediation; but experiencing is not yet understanding.

3.4.2 The World as Mechanism

The seventeenth century also bequeathed to us an entirely different approach to the explanation of the success of physics: the mechanistic world view. According to this view, the objects of nature are nothing but bodies. These bodies have certain objective properties, also referred to as primary qualities: extension, impenetrability, hardness, perhaps mass. They also have certain secondary qualities, which are actually merely subjective, such as colour and smell; the secondary properties appear only to our senses. A body we perceive as red and fragrant does not in itself possess these properties, but—in modern terms—emits light of a certain wavelength and sends out little molecules that trigger the sensation of a scent in our noses. Things in themselves are thus extended bodies and nothing else. These extended bodies are divisible. The mechanistic world view has usually gone hand in hand with the doctrine of atomism. In the last analysis, things consist of smallest parts, the atoms, which are not in turn further divisible; i.e., which, strictly speaking, no longer have any parts.

By thus reducing everything to the corporeality of things, one can perhaps hope to trace all forces occurring in nature back to the pressure and collision of bodies. Pressure and collision are simply the effects that impenetrable bodies exert on each other because of their impenetrability. It may be conceivable that the laws of mechanics could in this way be reduced to the necessary properties of impenetrable small bodies. This is the program of a mechanistic explanation of physics. Were it to succeed, the astonishing marvel of mathematical natural science would be reduced to a notion that to this mode of thinking appears a lesser marvel; namely, to the notion that all things are bodies.

---

[7] Cf. Heisenberg (1971a).

[8] Cf. "Parmenides and Quantum Theory", Chap. 8 in this volume, reprinted from C. F. von Weizsäcker: *The Unity of Nature* (New York: Farrar Straus Giroux, 1980), pp. 379–400; it was translated by Francis J. Zucker from: *Die Einheit der Natur* (Munich: Hanser, 1971): IV.6.

I will not discuss the unsolved internal problems of the mechanistic world view,[9] since this approach has become obsolete anyway. We do have an atomic physics today, to be sure. But according to our present knowledge, the chemical atoms are anything but small billiard balls, anything but extended lumps of matter. They can be separated into elementary particles that can in turn be transformed into one another; and the laws of transformation are described by a non-intuitive physics, for example by the mathematical formulas of quantum theory. The mathematical approach of the time of Kepler and Galileo has proven more durable than the attempt to reduce it to what was at the time called mechanism.

Historically, the mechanistic world view is closely connected with the relation between science and technology. Technology, especially in the seventeenth century, was largely a mechanical technology. Techniques for moving bodies by means of levers and pulleys had existed since antiquity. In the meantime, stimulated by the needs of the military, ballistics had come into existence. So had the art of shipbuilding. The mechanistic world view describes the ultimate realities of nature as if they were entities technically manipulable with the conceptual means of seventeenth-century technology. This is one of the occasions that show us how much a child of his time the scientist is in naively believing that his hypotheses express the plausible, perhaps even the true. The mechanistic world view would certainly not have been considered plausible in the world of the highly developed Greek science. It would have appeared totally alien to the majority of Asiatic cultures. How can one possibly set aside the immediate realities of mind and soul and in their stead base everything on something as low, as crudely organized, and as fundamentally uncertain as the mechanical properties of things? From the point of view of most Asiatic traditions, this is folly. To what extent, therefore, are the very foundations of science determined by historical situations, social interest, technical possibilities, and economic conditions? Once again we are faced with a problem that has no easy one-sided solution. One cannot simply say that Galilean physics was successful because it corresponded to the spirit of the times, which was bent on changing the world technologically. Surely it also succeeded because it was true, or because it contained at least a corner of the truth. In this manner, one *could* change the world, which a system of mere delusions could never accomplish.

3.4.3 Empiricism as a Description of Physics

Galileo's physics, as we saw, passed a test of reality, even though both its approach and its design were conditioned by the times and are today obsolete. But what is that: a test of reality? How was the test made? It was made experimentally.

---

[9] Cf. C. F. von Weizsäcker: *The Unity of Nature* (New York: Farrar Straus Giroux, 1980), pp. 101–137; it was translated by Francis J. Zucker from: *Die Einheit der Natur* (Munich: Hanser, 1971): II.1.

It was made empirically. In a countermove to the naive metaphysics of the mechanistic world view, a mode of thought now appears which is methodologically subtler by far, which incidentally also has ancient precursors, and which I will call empiricism for short. Empiricism sees the basis for the success of physics neither in our entering into the transcendence of the creative thoughts of God nor in entrusting ourselves to the more clandestine transcendence of the atoms or of other preconceived models of reality; it sees the reason precisely in our remaining in the here and now, in the palpable, in faithfully observing what one really experiences. Experience is the true and essential foundation of physics.

There is a very beautiful passage in John Locke's writings[10] where he says that, unlike the metaphysician, he does not wish to fathom the deep sea; the plumb line he and all of us have would not reach deep enough. He wishes only "to reach the bottom at such places as are necessary to direct his voyage, and caution him against running upon shoals that may ruin him". This thought can be expanded upon. A plumb line will in any case reach beyond the depth of the keel. For the practical purpose of a sea voyage, the plumb line one has is sufficient. We want to accurately observe what we encounter daily, and possibly also what we can technically produce; on such observations we want to base our physics.

What role does mathematics play in this empirical scheme? The empiricist does not deny that we can formulate the laws of nature in a mathematically simple way, and he must explain how this is done. Although a somewhat superficial empiricism had recourse to the invention of historical myths—for example, to the myth of a purely empirical Galileo—the newer, more sophisticated empiricism long ago started asking itself about the role of mathematics in empirical science. Its answer, in short, is: we formulate mathematical hypotheses and test them against experience.

That is how Galileo formulated the hypothesis that the distance traversed in free fall is proportional to the square of the elapsed time. He had good reasons for arriving at this hypothesis, and he then tested it empirically. Newton proposed the hypothesis of the general law of gravity, which he then tested for the case of planetary motion. From the antecedents of Newton's hypothesis I would like now to tell a story whose methodological relevance is not diminished by the fact that the authenticity of a particular anecdotal feature in it—the apple tree—cannot be vouched for.

Because of a plague, the young Newton fled from Cambridge, where he had been studying, to his home in the countryside of eastern England, and one day—so the story goes—was lying under an apple tree. He was a farmer's son; on this particular day he had probably been doing farm chores for as long as a theoretician could bear. An apple fell down and he asked: "What is it actually that pulls the apple down?" And he saw the moon and said to himself: "Why doesn't the moon fall down?" The answer is: the moon does not fall down because of the law of inertia, because it has a strong lateral motion. If it did not have this motion, the moon would fall down. Newton calculated the force that keeps the moon in a circular orbit around the earth. He compared the magnitude of this force of

[10] Locke (2008), Book One, chapter 1.

attraction with the force exerted by the earth on the apple, and postulated that the attraction varies in inverse proportion to the square of the distance from the centre of the earth. The result 'seemed to be' wrong. Newton saw that the formula was almost, but not precisely right. He let the matter go at that. For 10 or 15 years he did not bother about the problem.

Then came new measurements of the earth's diameter, and thereby of the distance to the moon, and with the new, improved data it turned out that Newton's calculation was exactly right. Only at that point did he dare to return to his old problem, which then led him to the formulation of the general law of gravity. What had happened? Newton had the correct thought, we might say the ingenious thought, from the first. But he was so deeply steeped in the empirical thinking of the new age that even a small discrepancy between experience and the deductions from his thought prompted him to reject the thought in its entirety. Later on it turned out that the thought had been right and the experience wrong. The experience had not been a good, an exact experience. Yet only when the exact experience had become available did Newton dare to believe his mathematical hypothesis.

This is how experience and mathematical hypothesis affect each other. When later on we speak of other sciences—for instance, biology or the social sciences, which try to discover general laws by empirical means—we shall have to constantly ask ourselves whether hypotheses can in fact be set up there, as they can in physics. A science is not empirical merely because it has collected a vast amount of material or because it somehow organizes this material mathematically in tabulations and statistics. In order to become a mathematical–empirical science in the sense of physics, this material would have to lend itself to a precise theoretical formulation that enables one to predict what will happen—and the predictions must then hold. The self-confidence of a science conceived in that sense shows itself in the ascetic attitude toward inexact confirmations, as the example of Newton amply shows.

3.4.4 Difficulties of Fundamental Empiricism

I distinguish between empiricism as a description and fundamental empiricism. The former describes what science—especially physics—is like, and in so doing gives experience its due. The latter is the attempt to fully explicate the possibility of physics in terms of the concept of experience; in other words, it is the claim that the concept of experience is sufficient to make us understand how physics can be possible. Empiricism as a description seems right to me; fundamental empiricism gets into difficulties which, I believe, stem from its attempt to explain an obscure idea by means of one that is equally obscure. What do we mean by the "testing of a hypothesis through experience"? Under what conditions is testing possible? Let me discuss some of the difficulties that arise.

Physicists usually put it this way: Newton confirmed his hypothesis empirically. He computed the planetary orbits, and the result corresponded to experience. The law of gravity is empirically correct. The curious phenomenon of progress in physics is met with also in its foundations, however. Physics not merely expands but also deepens its foundations. Thus, in the first quarter of our century, Einstein developed the so-called general theory of relativity, which deposed Newton's general mechanics as well as his law of gravity; merely as limiting cases do they still remain valid. Since Einstein all Newtonian formulas are, strictly speaking, wrong. They are approximations, good ones in most cases, but experimentally poor in others. How shall we put this?

If one demands the strictest application of the term 'true' to a theory, then it follows that Newton's theory was not true; although uncontested for two centuries, it was wrong. This way of putting it obviously does violence to the real state of affairs. After all, Newton's theory was nearly true. But what does 'nearly true' mean? In mathematics such an expression would hardly be admissible. What is this peculiarity of the empirical that allows us to 'nearly' describe empirical phenomena by means of mathematics, and then sometimes to describe the deviations again mathematically, but now by means of a new theory that replaces the old one?

But can one, by means of experience, confirm even one universal law? If I say, "All (healthy[11]) mammals are four-legged", I can confirm this empirically only by exhibiting all mammals so as to demonstrate that they all do have four legs. But how can I exhibit all mammals? Some are unknown, some have not yet been born. In the future mammals might evolve that are not four-legged. How can one empirically confirm an assertion that refers also to the future? In the eighteenth century, David Hume formulated this question with absolute clarity.[12] How can I know empirically that the sun will rise tomorrow? I will know this tomorrow, but then it no longer asserts anything about the future. One can contend that the proposition "the sun will rise again" has always stood the test. What has so often come true in the past can in good conscience be assumed to come true again in the future. This universal principle is sometimes called the principle of induction. How do I know that the principle of induction holds? Because it has met the test of experience, that's how. If the principle of induction is right, I may conclude that it will also be confirmed in the future; i.e., that the principle of induction is right.

It is simply impossible to in strict logic draw conclusions for the future from the past. Nevertheless, the whole point of physics is to conclude for the future. Physics predicts. It transforms the world. In the design stage, technology is always future technology. Empiricism, in other words, is incapable of elucidating an absolutely

[11] Here I have to make use of the 'norm of health'; cf. part 4 of "Models..." in Chap. 6 in this volume.

[12] Cf. "A Sketch of the Unity of Physics", Chap. 6, in: "*Major Works in Physics*" in this series, reprinted from C. F. von Weizsäcker: *The Unity of Nature* (New York: Farrar Straus Giroux, 1980), pp. 168–222; it was translated by Francis J. Zucker from: *Die Einheit der Natur* (Munich: Hanser, 1971): II.4.

basic feature of empirical science. As a description, it does say correct things, but it seems not to understand what it says.

Because this question is so important, I will discuss it in a few more of its guises.

Hume had already used the concept of probability. One can say: although it is not certain, it is nonetheless very probable that the sun will rise tomorrow. This concept of probability is ambivalent, however. One can take it subjectively. If I say, "this is probable", I imply that "I believe it is so." Subjective probability describes my psychological state or my behaviour. In this sense I indeed think it probable that the sun will rise tomorrow; I truly expect it. Hume cites a very good reason why this subjective probability in fact does come to pass: we have gotten used to certain regularities. Habit,[13] says Hume, explains our belief in the law of causality. This may be so, but what has it to do with the question of the justification for that belief? The question is, after all, whether the events to which I have become accustomed in the past will recur in the future. That I imagine it to be so today proves nothing, strictly speaking, about its coming to pass. Those subjective probabilities imply objective probabilities enabling one to predict percentages of future events can at best be true if one already knows that objective probabilities exist.[14]

In our time, Karl Popper gave the problem a somewhat different turn. A universal proposition ('all S are P') can be confirmed empirically only by exhibiting all instances. Since this is practically impossible in experience, one might claim that, in principle, experience cannot verify a natural law stated in the logical form of a universal proposition. What does empirical science accomplish, then? It formulates universal propositions that are useful so long as they have not been falsified. A single counter-example will falsify a universal proposition. The empirical character of a science shows precisely in the falsifiability of its propositions. An assertion is unscientific if formulated in a manner that prevents our stating what must be the case if the assertion is to be empirically falsified. This gives rise to a critique of psychoanalysis and Marxism, insofar as each of them claims to be a science.

It seems to me that Popper's assertions clarify empiricism as a description. We indeed do work with hypotheses until they are falsified. But if his assertion is taken in the sense of a precise formulation of what I have referred to as fundamental empiricism, then two objections come to mind.

The first: the great amazement, which I tried to formulate as the amazement of the physicist at the fact that nature can be described mathematically, is not explained, it seems to me, by saying that hypotheses are put to use until falsified. The great marvel is the very existence of universal propositions that can be relied

[13] Hume used the term 'custom'.—*Translator.*

[14] Cf. "Quantum Theory", Chap. 7, in: Michael Drieschner (Ed.): *Carl Friedrich von Weizsäcker: Major Texts in Physics* (Cham et al.: Springer-Verlag, 2014); reprinted from C. F. von Weizsäcker: *The Unity of Nature* (New York: Farrar Straus Giroux, 1980), pp. 181–222; it was translated by Francis J. Zucker from: *Die Einheit der Natur* (Munich: Hanser, 1971): II.5.

on for quite a while without their being falsified. This is no way of understanding how solar eclipses can be computable over millennia. Counter-examples show how amazing a phenomenon this is: it is not met with in meteorology, or in the social sciences. Why is it, then, that those marvels exist at all?

The second objection is that one cannot even falsify a universal proposition without presupposing the verification of certain other propositions. Let me once more call to mind the example of the refutation of the energy law. A universal proposition, the Law of the Conservation of Energy, was seemingly falsified by a counter-example, by a measurement. But that counter-example was no counter-example. It had to be produced by means of fairly complicated apparatus, which in turn had to be checked out with the help of a theory of that apparatus. This theory applies universal laws of nature to the apparatus, and here is where the author of that paper made a mistake. The mistake was so well hidden that he failed to discover it. Some good theoreticians saw that there must be a mistake somewhere, the result "just could not be correct". At this point, I must admit, the anecdote does not accurately match the intent of the objection. The objection, raised by the logic of science, refers not to the avoidance of mistakes, but to the fact that a falsification can be no more trustworthy than the propositions which it presupposes as unfalsified. The seemingly profound difference in terms of certainty between verification and falsification is hereby cancelled.

3.4.5 Scientific Perception

The mistake of fundamental empiricism appears to lie in its vain search for certainty. The universal cannot be confirmed with certainty. It is not even true that experience is certain in relation to the individual case. What we found acceptable in Popper's point of view sooner points to the opposite solution: experience leads us to the universal, but not with certainty. In a sense, we perceive also the universal in the individual case.[15]

Assuming this assertion is correct, we still have not solved the philosophical question as to how it can be so. What must nature, what must the human perceptual faculties be like so that the individual case can 'let us see' the universal? And how do we now react to Hume's problem? I will have to set aside these large questions. It is more urgent to closely watch what happens in this perception of the universal in the individual, in this 'scientific perception'. We are in need of a psychology or phenomenology of scientific perception. To this topic I can today contribute only a few scattered comments.

In communicating a common perception—e.g., "this table is round"—by means of language, I have already expressed it conceptually. A concept refers to

---

[15] Cf. "Models...", Chap. 6, part e. of this volume; "Parmenides and the Greylag Goose", Chap. 7 in this volume.

something universal. If we can express it at all, the individual is already given to us in the medium of the universal. One could call it perception of structure, or predicative perception. This human faculty is the building material of scientific knowledge.

What characterizes good, productive science? First of all, I should think, a greater ability to perceive patterns, to ferret out especially-simple patterns that are the better concealed because of their simplicity. Let me illustrate this with another anecdote.

When as a young theoretical physicist I was a student of Heisenberg's, he assigned me problems to work out. I also devised problems of my own and discussed with him my ideas on how to solve them—that's where my doctoral thesis took its start—and in doing so I had to practice the craft of theoretical physics. I had to solve certain equations, perform certain calculations, and when I had computed a result I went to Heisenberg and showed it to him. Heisenberg didn't even glance at the first page, or at any part of the beginning for that matter; he looked only at the final result, thought for a while, and said: "That's wrong." And so I said: "Yes, but how come?" And he: "Well, no—it can't be like that." Proceeding from the end, he then checked the calculations until he found the mistake. He was almost unerring in this instinct. This is science. This is what goes on. This is what training in science is like. The true training of the scientist consists in one's being refuted by one's teacher over a period of several years, in learning to see again and again in what sense one was wrong. To be sure, one sometimes refutes one's teacher; not authority, but the development of perception is the issue in this process. I recall what Heisenberg, long after he had become a full professor at Leipzig and a Nobel laureate, told me about his revered teacher Bohr: "It used to be that Bohr was always right when we disagreed—today I am right in 30 percent of the cases." This is what goes on even in research. The truly productive, truly eminent researcher is distinguished by an instinct for, a feel for, a not quite analysable perception of interrelations that is deeper than that of most other people; this accounts for his being the first to arrive at a particular truth. A scientific truth is almost always first surmised, then asserted, then fought over, and then proven. This is essential, it is the nature of science, it cannot be any different. Later on, this truth perhaps becomes classical, then seemingly trivial, then someone discovers that it is problematic, and finally it is replaced. He who replaces it with a new surmised, asserted, fought-over, proven truth usually recovers the sensitivity for the ingenuity, for the non-obviousness of the knowledge he overturned.

And what distinguishes the truths that signify radically new progress? What is the criterion employed in this perception of structure? If you asked Heisenberg, he would gladly answer: "Nature is mathematically simple." The theories become progressively more abstract, to be sure, but to anyone who understands this abstractness it appears as a higher-level simplicity. In principle, the theories become progressively simpler. What is very simple can no longer be expressed in the form of concrete details, concrete images, because the concrete is always complex. The simplicity and abstractness of our modern theories are two different aspects of one and the same essential feature.

If you pressed Heisenberg further by asking: "What does 'mathematically simple' mean?" he could be brought to answer: "What is beautiful." With what right are aesthetic categories being introduced here?[16] Let us take this seriously as a perception also. First of all, Heisenberg is here accounting to himself, at a high level of methodological awareness, for his inability to account for the basis of this simplicity. Secondly, he reminds us, probably justifiably so, that the basic theories have something in common with great works of art. He says, and from the phenomenological point of view rightly so, that they are beautiful. Anyone who wishes to know what this beauty is would probably have to also know what beauty is in art. It may be that, as philosophical disciplines, the theory of scientific knowledge and aesthetics share a common ground in a doctrine of *poiesis*, the fashioning of patterns. But now I am at the outer limits of what we know, perhaps even a bit beyond them.

References

Gaiser, K., 1971: *Platons ungeschriebene Lehre* (Stuttgart: Ernst Klett, 1963).

Heisenberg, W., 1971a: *Physics and Beyond: Encounters and Conversations*, Vol. 42 (New York: Harper & Row; World Perspectives).

Heisenberg, W., 1971b: "Die Bedeutung des Schönen in der exakten Naturwissenschaft", (Stuttgart: Belser-Presse) (German and English), in: *Physikalische Blätter*, 27, 97–107.

Locke, John, 2008: *An Essay Concerning Human Understanding* (Oxford).

von Weizsäcker, C. F., 1964: *The Relevance of Science—Creation and Cosmogony* (New York: Harper & Row): 54–76.

[16] Cf. Heisenberg (1971b: 97–107).

Carl Friedrich von Weizsäcker discussing with Erhard Scheibe; in the background Wilhelm Ochs. (Conference on Quantum Logic, Cologne 1984). © Hans Berner who granted permission for its use in this volume

Chapter 4
Time: Physics—Metaphysics

I have been led into temptation and have succumbed to the temptation to sketch the complete outline of a philosophy in somewhat over an hour.[1]

This presentation will have four parts. They pursue the concepts mentioned in the title—time, physics, metaphysics—in reverse order; and thus we come full circle. I begin with a glance at the history of European *metaphysics*, which tradition my outline follows. I shall continue with the philosophical interpretation of modern *physics* in the form of quantum theory. Both paths lead to the central problem of *time*. I shall pursue this problem initially within the context of natural science under the title *evolution*. Evolution broadens out into the history of human *culture*; and the role of *metaphysics* emerges within the context of European culture.

4.1 Metaphysics

Within the tradition of Western philosophy, metaphysics means what comes after physics, with reference to the title given to a collection of essays by Aristotle. Physics here denotes the knowledge about that which becomes and comes to pass, a motion; with Aristotle, physics is distinguished from ethics and poietics which concern human dealings and doings. After physics comes the question of what must be the case in order to make physics possible at all. To that extent, metaphysics is reflecting about physics. In a similar sense, authors of our century then speak of meta-logic or meta-mathematics. Just this division between knowledge and meta-knowledge is problematic, of course.

[1] This text is taken from: *Der Mensch in seiner Geschichte* (Munich: Hanser, 1991): 86–109. It was translated for this volume by Ms. Ann Hentschel with the financial support of the Udo Keller Foundation.

M. Drieschner (ed.), *Carl Friedrich von Weizsäcker: Major Texts in Philosophy*, SpringerBriefs on Pioneers in Science and Practice 23,
DOI: 10.1007/978-3-319-03671-7_4,

In the preliminary historical commentary I will fleetingly stop by five stations of Western philosophy, signposted by the names Plato, Aristotle, Kant, Hegel and Heidegger, with a glimpse at the Indians.

Plato expounds the problem in the dialogue *Parmenides*. Things that have a becoming and coming to pass have their being through participation in the idea. But what is this idea? Among the four transcendentals, the first is treated here: The idea is *one*. The being of the one and of the many is the topic of the dialogue. Its exposition is aporetic: This being is not conceivable by means of logic. Only within the range of the many is there negation and by its negation, affirmation, hence logic. The aporia is that the many cannot be without participating in the one; however, the one cannot be conceived according to logic, which sunders being from not-being. That is why the argument is entrusted to Parmenides; the key point of the dialogue is shown in that Plato allowed only him, someone greater than Socrates, to speak. The tension between the one and the many is not played out between idea and phenomenon anymore but rather inside the idea. Plato calls knowledge about the idea itself dialectic. That is why this dialogue is dialectic. Here is a link to Hegel's usage of the word dialectics. Hegel emphasizes as the essence the progress of a thought through what—according to logic—is contradiction. He reads the dialogue as the coded program of a systematic philosophy, which I would like to call the philosophy of the descent into the cave. The thought thinks the 'actual being', therefore the progress of the thought, i.e., the motion of the thought, thinks the 'actual-being' motion. Motion is thus the key concept of the mediation between the one and the many. Plato's philosophy of descent, i.e., his actual philosophy, is a philosophy of motion.

The decisive divergence of Aristotle from Plato lies in his contesting the inapplicability of logic on the one and on being. Insofar as logic is the adequate regulator of thought about actual being, Aristotle thus creates the form of philosophy that Heidegger terminologically denotes as metaphysics, in which being is treated like 'actual being'; the ontological difference is not adequately conceived. Heidegger's resurrection of the issue of being is, to this extent, a resumption of Plato's issue against Aristotle. Heidegger himself could not see it this way, because Plato's philosophy of descent had to seem to him as an anticipation of mathematical physics—and a necessarily immature one, at that—therefore, as an anticipation of a form of metaphysics that in Heidegger's meaning was remoter still from the source. This reveals a limitation of Heidegger's revisiting of Greek philosophy: his alienation from the central motive for this philosophy, the discovery of mathematics. On the other hand, this motive cannot be appreciated as long as mathematics itself is not philosophically understood. In any event, the way Aristotle encounters the difficulty of thinking of being is that he can only lend a kind of unity to his enumeration of multiple ways of speaking about actual being through the intransparent thought of analogy. Analogy equal to proportionality is a mathematical term that is used metaphorically in philosophy.

A philosopher schooled in the West can best meet Indians along the path of Neoplatonism. While conversing with Sanskrit scholars in the tradition of Kashmiri Shaivism in Srinagar, it seemed astounding that I was able to classify somewhat a priori the—to me unfamiliar—conflicting schools of the Vedanta. My guiding thread was the positions of the Platonic Parmenides dialogue. In discerning observation of bodily-psychic, mental and supramental experience, the Indian tradition is far superior to the European one. What is lacking in Indian thought, in order for it to be philosophy in the sense we use the word, is the constitutive role of mathematics. I use this glimpse at Indians to introduce the concept of sat-chit-ananda, being-consciousness-bliss. The one is the identity of these three. A discrimination between the three only exists in the world of plurality. An individual concept is only meaningful where its negation can also be experienced: Being where there is non-being, hence for concepts suitable or unsuitable for a thing; consciousness where actual being remains unknown or unconscious of itself; bliss where suffering, unhappiness is experienced. This corresponds exactly to the Platonic transcendentals. The one is, first of all, being, which lends the many (ideas and derived therefrom perceptible things, which are actually plurally mobile ideas) its logically negatable actual beingness. To this same extent, it is even beyond actual beingness. The one is, secondly, truth, which casts revelational light on the many. The one is, finally and additionally, the good, the bliss that permeates everything and every human action, to the extent that thing and action suggest their ideas, and which is the trait of the *unio mystica*.

Kant has to be read in the present context as a critic of Western metaphysics following Aristotle, particularly that of modernity. Kant's epistemology about finite consciousness dependent on experience comprehends the unity of this consciousness as the condition for its possibility. The medium of plurality, which with Plato could be called motion, is called time with Kant, more generally: form of apperception. The possibility of unity in plurality, that is, the possibility of cognitive consciousness is, according to Kant, mediated precisely in that he understands time as a form of apperception of the consciousness, as constituting the subject. Understanding is the means of concepts; reason the means of unity (totality: unity in plurality). Our form of unity is the unity of sensation, time. The unity of the world in itself, of the person of himself, of the grounds of both in themselves, must be presupposed but is not a subject of experience. That is why it cannot be the subject of theoretical knowledge, even though the possibility of theoretical knowledge without it must seem incomprehensible; it is a regulative or moral postulate of reason, thinkable but not knowable. Dogmatic metaphysics, which does not see this limit to its capacity, deteriorates into a dialectical semblance. Dialectic is meant here in Plato's usage, as sought knowledge of the principles, at the same time as in Aristotle's usage, as field (and revelation) of fallacies. This semblance is transcendental, i.e., necessary by nature of reason, even though reason can see through it. Kant thus hits upon just those problems that had led the synthesist Plato beyond logic, the analyst Aristotle to content himself with the analogy concept—problems not set clearly enough in view for medieval-modern metaphysics, still naïvely grown out of the Greek school.

Hegel—instructed by Fichte and Schelling—sets the inconsistency of the Kantian doctrine of postulates straight. That which cannot be known, cannot even be thought. Well, we can think of the one—Hegel says: the absolute—, for without it we cannot think anything. Therefore we can know it. However, the motion of contradiction changes all concepts in that it effectuates them. When we know the absolute, we know ourselves as forms in which the absolute has clothed itself. In ourselves the absolute knows itself, for the absolute is the sole subject. This philosophy once again performs, by its intent, Plato's ascent and descent, both as motion through contradiction. It being a modern-day egalitarian, it does not remain an unwritten doctrine hidden behind aporia and myth of the dialogues; rather it criticizes public affairs publicly and thus itself confronts the public critique. This critique provoked by Hegel had to turn out scathing on many points; his systematic claim counts as obsolete since the middle of the nineteenth century. Whoever uses Hegel's work since then as a quarry like a ruined castle, as more recently many are doing, may find use in it but fails to recognize that the gist of Hegel's thinking lies precisely in his system of thought. Dialectics is the recognized impossibility to grasp the identity of the singular, because the singular when it is recognized as what it is, proves to be conditional upon the relevant truth, which is the whole. It is the presence of the whole within the singular, of the one within the many. If we of today now ask whether we can follow the reflection taking place in each step of Hegel's dialectic, we should initially have to say, in psychological self-observation: Yes, as a process in time. First we think of a concept, thereafter we ask what it meant. By our finally also asking about the basis of truth in the logic and finding it in the rules of acting, thus of the behaviour in time, time becomes for us the precondition for the concept as well, differing from the metaphysical tradition, which at best Kant's observation affronts here but not his theory.

Heidegger takes just this radical step, by intention in any case, to full measure. Time appears to him as the horizon of being. The important thing about it is that this thought is not just expressed in isolation but is grounded in an examination of the history of Western metaphysics altogether. Heidegger sees this history not in one of the two common superficial forms, either as a chain of advances or as an error to be surmounted. Rather, he sees in the received phenomenological aporetic of the Greeks—not in its problems but just in its solution models—the guiding thread along which metaphysics, up to theoretical physics of our century, moved with a necessity concealed even to itself. If this is correct, the step due today presupposes just this knowledge of time as the horizon of being, which is a debate not foremost with the forms of modern metaphysics but with its Greek origin. This debate Heidegger, in my view, only introduced, however; he did not engage in it. I do not presume to understand the Greeks better than Heidegger. But I must emphasize one point, that Greek philosophy—that is, philosophy—without the key role of mathematics must remain incomprehensible. Among the philosophers after Plato whom I have discussed here, only Kant saw this; among those not discussed here, surely only Descartes and, more profoundly, Leibniz, and then the school beginning with Frege. What, however, remains concealed to the philosophers who did take mathematics seriously—with the exception of Kant and Brouwer—is the central importance of time in mathematics.

This is how the history of philosophy appears to me to have staked out the field of today's tournament.

4.2 Physics

The unity of science is not located in its method but in the content of its central theories. Of philosophical relevance in current science is consequently not theoretical philosophy of science but the substance of the science itself.

Logical positivism wanted to be a philosophy of science that itself was a science. It legitimately saw the concept of experience as central to science. If positivism had itself been a science by its own definition, it would have had to learn from experience what experience is. This radical step was first taken by Thomas Kuhn by transforming theoretical philosophy of science into history of science. His most important finding was: Science is not a continuous accumulation of knowledge but an alternation between plateaus and crises. The plateaus Kuhn calls normal science, the solving of individual problems under a firm unquestioned paradigm. The crises are the scientific revolutions, the paradigm changes. Philosophizing, a mountaineering art, is hardly applicable on the plateau, but indispensable during crisis.

Physics is the central discipline of science. Today we know no limits to its range of validity. Heisenberg said already 15 years before Kuhn, in 1948, what a revolution in physics is: the transition from one closed theory to another more comprehensive one. Heisenberg calls a theory closed that cannot be improved anymore by minor alterations. Just for that reason it can only be replaced by a revolution. Newton's mechanics, Maxwell's electrodynamics, Einstein's special and general theories of relativity, the quantum theory developed by Planck, Bohr and Heisenberg are closed theories.

The phenomenon of closed theories deserves utmost admiration. The basic assumptions of quantum theory can be described for a mathematically educated reader on a single printed page. The number of individual known observations obeying quantum theory may well amount to a billion today; and not a single experience has been found that would credibly contradict quantum theory. I employ a thought by Kant to suspect that quantum theory is generally valid in experience because it formulates conditions of possibility for experience.

At this place I can only explain this by *one*, obviously, the central topic, the importance of *time*. In doing so I do not proceed by systematic deduction, but along the lines of a propaedeutic philosophy of ascent. I shall portray how the role of time in physics became visible to me personally, step by step.

It began with the irreversibility of physical processes. As a pupil I asked myself: Why do unlucky train accidents exist in which an intact train comes suddenly to a halt and is destroyed, but there are no railroad "strokes of luck" in which a destroyed train suddenly sets off and is intact? As a physics student I learned that the Second Law of Thermodynamics describes irreversibility: The entropy of a closed system can grow but not diminish. Statistical mechanics explains this law. The entropy of a state is a measure of the probability of its occurrence; more improbable states pass over into more probable ones. I wrote a paper about this as a young researcher that I still consider correct. How can the concept of probability distinguish later events from earlier ones? How, therefore, can it define what physicists metaphorically call the direction of time or, even more metaphoric-ally, the arrow of time? The answer is: Every person of our civilization, indeed, every physicist implicitly always understands the difference between past and future. What has passed is factual; it has unalterably happened. What is to come is possible. Probability is a quantitative, mathematized version of possibility. The probability of events in this direct sense in which physics uses it here, thus always refers to the future. The qualitative distinction between the past and the future is not, as physicists sometimes conjecture, a consequence of the Second Law. It is rather its phenomenological premise. Only because we understand it right off can we conduct physics the way we do.

I later saw that this distinction is constitutive for the basic concept of experience. Experience means learning out of the past for the future. Time in the sense of this qualitative distinction between fact and possibility is a condition for the possibility of experience.

I tried to understand quantum theory along this guiding thread. Quantum theory is a statistical, a probability theory. Its core is a nonclassical probability calculus, characterized by the so-called superposition principle. Probability calculus presupposes logic. Nonclassical probability calculus of quantum theory presupposes, according to J. von Neumann, a nonclassical logic, which he called quantum logic. It has been argued that all experience already presupposes the one true logic that we call classical. So it would be absurd first to base a special theory, such as quantum theory, on experience, and only afterwards to distil out of quantum theory a diverging logic. This objection is, in my opinion, strong but false. Experience presupposes time. The logic by which we describe the laws of experience must be a logic of temporal statements. Predictions, statements about the future express possibilities. In logic, possibilities are called modalities. I tried to sketch a logic of future modalities. A link can be made here with the foundations of mathematics, incidentally. Intuitionism or constructivism initiated by Brouwer regards mathematics as an intellectual act; I like to say: a generation of intellectual forms. This happens in time, as Brouwer has pointed out. The fundamental logic for this must be temporal logic. But now I stay with quantum theory. I believe I have shown in my papers that quantum logic can be understood as a special version of temporal logic. Within it, logical structures would then occur to our consciousness that in truth are already the basis of all experience.

Now I shall make a brief remark about the content of quantum theory that within the confines of this presentation can only be comprehensible to physicists. The q-function, also called state vector, has, as we have seen, a future sense. It defines the probabilities of all possible measurement results for the relevant object. Now, measurement results must, according to Bohr, be described by classical concepts. Why? I subscribe to the interpretation that here the word 'classical' stands for 'irreversible'. Measurement results, once they have been gathered, are facts; then they have unalterably happened. The profound problem of quantum theory, never yet sufficiently understood philosophically is, however, that owing to the superposition principle any classical, factual description of an event is just an approximation. If quantum theory is right, then reality is, strictly speaking, never factual.

I explain this by the composition rule of quantum theory. The state space of a composite object is the tensor product of the state spaces of its parts. In this total state space there is only a set of measure zero of such states in which the component objects are in well-defined states at all. This is the end of the comment for physicists.

I philosophically conclude: If quantum theory is right, then, as I said, reality never is rigorously factual. If current physics is right, the object concept itself, upon which this particular physics rests, is just an approximation.

But we know now of no limits to the validity of precisely this particular physics, quantum theory. It seems to encompass all inorganic processes. Even in biology, physicalism—the doctrine that physics is the sole, sufficient fundamental discipline—is steadily successful today. I personally feel absolutely no need to doubt physicalism. The question then arises whether quantum theory is also applicable to psychic processes, also to the consciousness. My hypothetical response is: yes. One has to ask what such a hypothesis signifies. Quantum theory, formulated entirely generally, that is, abstractly, makes no presuppositions about the content as to its objects having to be objects in space. It is a theory of probability predictions for arbitrary decidable alternatives. Now, there is an approximation, in which the question of the state of consciousness I shall be in tomorrow morning—e.g., cheerful or sad—can be made into a somehow decidable formulatable alternative, such as, through introspection. In this approximation quantum theory ought to be applicable to the consciousness. My argument thus, as you see, does not take the 'materialistic' detour. It does not say: "Quantum theory is applicable to the brain and thinking is a function of the brain." The argument is direct: "Insofar as my state of consciousness can be the object of a logically formulatable and factually decidable alternative, it will satisfy the abstract theory of *all* predictions for alternatives, specifically of quantum theory."

I confess that I never did understand the Cartesian dualism between thought and extension and that I deem it the consequence of an error in the logic. He stylizes a difference between two roles in the act of cognition that of the cognizing subject and the cognized object, into a distinction between two substances. To Descartes, extended means that which is accessible to the only clear and distinct intelligence which is not self-consciousness. This for him is mathematical—more specifically—geometric knowledge. To us mathematics is more abstract. Quantum theory is our theory of mathematicable predictions. To the extent that I myself can become the object of cognition, I should thus fall under quantum theory. I add, of course, that perhaps my consciousness itself—insofar as it satisfies quantum theory in its approximation which we call classical—must become perceptible as an extended object (e.g., as a brain).

I have last of all wandered into the field of hypotheses. The sheer logical possibility of these hypotheses shows, however, that quantum theory would be easily reconcilable with a spiritualistic, monistic metaphysics. If we are philosophers, our first question must obviously not be whether we want to believe such a metaphysics. The question must rather be whether we know what we mean when we contend that such a metaphysics is conceivable.

I call an object finite when it can be characterized by finite or countable infinite alternatives. I call a cognizing subject finite if in its cognition it is limited to finite objects. Then quantum theory would be the most comprehensive theory of finite objects known to us. Hence it would be the current science of how the world must appear to finite subjects. How do we ourselves figure in this world?

4.3 Evolution

We are children of nature. Nature is older than man. But man is older than science. This fact allows our philosophy a circular path that classical metaphysics could not take. Classical philosophy knew of man as knower, of nature as the known. But it did not know of the natural history of human knowledge. Classical philosophy shut itself off from the productivity of thought in such circles because it had formed according to the paradigm of deductive mathematics: It sought unquestionable beginnings and necessary conclusions, even though the ascent to the beginnings was the hardest work. We, however, now seek the circular path. We seek the mutual commentary on nature through knowledge and on knowledge through the history of nature. This circular path of philosophy, too, is a motion in time.

In this description of evolution I basically follow the Darwinian theory, in particular some considerations by Konrad Lorenz and Karl Popper.

Evolution, indeed life generally, is gnoseomorphic, knowledge-formed.[2] This means, firstly: evolution accumulates information. I shan't now go into the

[2] "gnoseomorphic character" for German: "Erkenntnisförmigkeit".

definition of the concept being used here, of objective information. I only note that the preponderant probability of this information growth, if one argues exactly, does not at all contradict the Second Law of Thermodynamics; but the former just as the latter are themselves a consequence of the probability concept, if it is temporally interpreted. Evolution is a form of entropy growth (I am now not discussing the equivocations connected with the sign of entropy. In short, I say: Entropy is virtual information).

Measurable information is always defined under a concept. It is larger, the more encompassing, hence the vaguer the concept is. The information of a telegram is lesser under the concept 'message in the French language', greater under the concept 'series of Latin letters', greater still under the concept 'series of signs'. We call information measured this way objective. It is information to us researchers and defined by the concept such that, up to the individual assessment spread, every researcher will obtain the same number. Subjective information is what we would call it, when we read it off the behaviour of a single individual.

Objective information is, for example, information of a set of chromosomes under the concept 'genotype'. Animal behaviour, on the contrary, is an evolutionary level upon which for the first time there is something we could call subjective information: the informational content of the stimulus for an animal, which we researchers can read off as the animal's reaction. This is the first preliminary stage of animal knowledge.

The simplest behavioural schemes can be characterized by four traits that I shall call: generality, yes/no principle, prevalence of the positive, adaptation.

Generality: The animal reacts, in a way that to us is conceptually describable, to a stimulus that to us is conceptually describable. It reacts typically to this type of event. The marten, which kills all the hens in the henhouse even though he can only eat one, responds—antiquely put—to the eidos 'fluttering hen', not to 'this hen'. Exactly this is to be expected cybernetically. The simpler the computer, the more typified and unindividualized the stimulus must be for it to be able to make it out.

Yes/no principle: The simplest response schemes do what is required only when, depending on the stimulus, they are either on or off, without complex intermediate stages. The computer has a bivalent logic.

Prevalence of the positive: The somewhat more differentiated behavioural scheme responds specifically to rare occurrences; it sets them apart from the chaos of stimuli. In logic, likewise, the truth of a special, positive statement is a rare, improbable occurrence. The negation of an interesting statement, obversely, denotes uninteresting normality: "The world is full of not-elephants" (Bochenski).

Adaptation: The correctness of the behavioural scheme is its adaptedness to the living conditions. This adaptation is not one of a copy to the original image but that of a key to its lock. In logic this corresponds to pragmatic theory of truth. The truth of a judgment is the adequacy of the thought to the facts. I read this pragmatically: The truth of a judgment is the adaptability to the circumstances of the possible forms of action represented in the judgment.

I have just made use of knowledge-formed animal behaviour to discuss behaviour through knowledge and knowledge through behaviour. There are, hence, biological preliminaries to logic.[3] I now pick out only one: Plato and Aristotle were entirely right that one can only know the eidos. At least, the simplest animals can only do this. Man, especially in our highly developed civilization, can obviously make the complex achievement of distinguishing the individual case from the concepts under which it falls. Erroneously, nominalistic philosophy then holds the individual case as the elementary datum. Presumably there is no such thing as elementary data.

What, though, does this behaviour describable in computer terms have to do with consciousness?

Chuangtse and Huitse were walking along the riverside. Chuangtse said: "See how the fishes are leaping out of the water! That's the fishes' joy." Huitse interjected: "You aren't the fishes. How can you know that this is the fishes' joy?" Chuangtse replied: "You aren't me. How can you know that I do not know that this is the fishes' joy?" And after another exchange of logical blows Chuangtse concluded: "I know the fishes' joy from my joy at looking on."

I already anticipated this reversal of scepticism as I was speaking about quantum theory. To us animals and computers are composed of atoms. Along the by-road of physics we distinguish finite alternatives on the behaviour of animals and computers. What are the animals of themselves?

The consciousness emerges over the course of evolution out of the sea of the living. Evolutionary theory taught us that stages, evolutive plateaus, do not signify distinctions in substance. Only the human consciousness capable of self-reference, of reflection can pose the question: "What is consciousness?," and likewise the questions: "What is an object?", "What is a living being?" The consciousness asking thus, knows itself today as an evolutive stage, delimited from older stages by achievements that according to our knowledge require a definite level of complexity of the organic apparatus—mind you, a complexity level that our present-day computers do not possess! Processes such as perception, emotion, action, that we can set apart in our own behaviour through judgmental reflection, rather form a continuum in animal behaviour. We have trouble imagining this inseparableness of theirs as a psychic state. The concept of the psyche, of experience itself, is a product of the level of consciousness.

We help ourselves along in that we consciously picture for ourselves the earlier stages as potentially conscious, as evolutively designed toward the ability of consciousness. All being seems to be connected with the consciousness by a historical continuum; said the Indian way, all 'sat' would be potential 'chit.' All organic behaviour at all understandable to us is emotively understandable; emotion-free behaviour is a human artifact. That means, all living behaviour

[3] Cf. Chap. 5 in this volume.

understandable to us 'strives after bliss', is potential 'ananda'. This is the lesson of the fishes' joy. Should we say that atoms also are potentially conscious and yearning for bliss? But atoms are finite objects. The atom is a concept of human potential action. Presumably atoms exist just for us. Perhaps just for us physicists at the end of the second millennium of the Christian Era. We must walk full circle down the path, presumably many times. So, now we must take a look at the cultural premises of our science.

4.4 Culture

We are philosophizing *now*. We are not philosophizing in eternity. We are talking about time, in time. In the midst of the technical age we are asking about the origin and sense of concepts that this age impresses upon us.

I have tried to gather together a few concepts into a loose structure, which I shall call a scaffolding. A scaffolding has no firm foundations. It clings to the building that is being constructed by its assistance. And in the end it is supposed to be torn down again. The building would be a historical understanding of man. The scaffolding consists up to now of four stories, counted upwards from below: (1) the unity of perception and motion; (2) the frame of purposive rationality; (3) the triad of modern cultural summits: theory, morality, and art; (4) the unity of truth. Thus formally the frame rests on unity and culminates in unity; in between there are a plurality of levels.

The unity of perception and motion, of stimulus and behaviour is of animal heritage; I have just portrayed its differentiated structure. In man one distinguishes four moments: perception, judgment, emotion, action. I repeat an old example. The car driver *sees* the approaching red contour, *judges*: "a car in the wrong lane", *takes fright*, and *acts* by swerving past it. In their normal course these moments are unseparated. Perception is predicative: I see the car (I see the eidos!). The judgment is emotive: "Yikes!, on the wrong lane"—even "wrong" is an emotive concept. Perception, judgment, emotion lead (if they succeed) seamlessly to the action.

How do we know, though, that our conduct contains these particular four moments? One part of the answer is given by a glance at the level immediately above, the level of utility, or as the learned say, purposive rationality. The floating nature of the 'scaffolding' is shown in that always only the higher level explains the structure of the next lower one, at least partly. We are rather going to assign the lower levels to our notions of the nature of man, the higher ones rather to the impressions of culture become historical. But then the 'floating' character of the scaffolding means that our notions of the nature of man are, in truth, themselves the product of our culture become historical. Exactly this fact is what my exposition is aiming at.

Within the thought scheme of purposive rationality, we juxtapose the judgment about the facts, unladen by action, against the action that we decide on for the sake of a given purpose. Here the concepts return that we used in the analysis of science. The judgment is assertive: It establishes a fact. The action can be thought and willed beforehand: It is reached within a scope of possibilities. Judging without acting is a kind of asceticism, an interruption of the natural course of reaction. Just this asceticism makes man capable—not of reacting, but—of acting. The experience of freedom is rooted within it.

Thus, however, we obtain only incomplete information about the four moments of behaviour. By casting light on the pair judgment/action, perception and emotion retreat into the shadows. Perception remains an unclarified residual category; that is why neopositivist philosophy of science at its beginning stage could fall for the idea that sensory perceptions were elementary data. In reality sensory perception itself is a highly complex achievement. Emotion, too, is an unelucidated residual category. Judgments are debatable; prescriptions can be set for actions. The affect, the emotions remain in the shadows that our culture likes to exclude from the discussion by employing the word 'subjective'. Whoever calls someone else 'emotional' almost denies him communication in today's usage—and rarely is a word itself so bitterly emotion-laden as the word 'emotion'.

One limitation of the thought scheme of purposive rationality is that the purposes themselves are not understood either but are rather presumed to be 'subjective'. One then speaks of 'values', for instance. And if one sees that all so-called values remain unfounded, that our concepts of value lack grounding, that judgment and action, understanding and will on their own ultimately only teach us will as an end in itself, then one speaks of the unsolved problem of meaning.

Our culture obviously does know grand values that stand above the level of utility. I mentioned three areas of such values earlier. I spoke of theory, morality and art. The conventional names of their guiding values are the true, the good, and the beautiful. Instead of morality I ought to better have used the Aristotelian term of practice, for the sake of those philosophically trained. Yet, Aristotelian poiesis covers besides art also technical engineering, which we would rather order under purposive rationality, therefore, under the value of utility.

I mentioned theory, morality and art as modern cultural summits. Modern, that is, not self-evident, not to be derived out of an imputed nature of humans. Rather culturally: artificial products. Summits: drawing together of a broad base of phenomena, each to a narrow pinnacle of achievements, three Eiffel towers.

This triad is, as I said, not to be derived out of human nature. I know of no persuasive systematic foundation for it. It rather seems to me to consist of a guiding summit, namely theory, and its compensatory remainder summits, first practice, then art.

Theory is, I feel, the genuine work of art of Western culture, which distinguishes it from all other cultures. Indian and Japanese persons have told me: "The strength and limits of your culture is the dominance of Aristotelian logic." My current presentation treats of theory, makes it the topic. We first encountered it as the deductive mathematics of the Greeks and as its sister development metaphysics. In modernity, science developed fully, to which mathematics is not just a paradigm but an instrument. The humanities and social sciences have orientated themselves according to science, to set themselves apart from it or to apply it. The core concept of theory is the true, which opposes the false: bivalent logic.

We started on a circular path, to study scientifically, namely, by evolutionary theory, the biological preliminaries of logic[4]: a beginning of a substantive theory of the theory. In this biological prelude we learn to appreciate the functional utility of the yes/no decisions; one can also say: their power-formedness. Precisely from there, however, we learn to be sceptical about its metaphysical truth. Konrad Lorenz and Karl Popper, though, who denote their philosophy as 'realism', think only behaviour adapted to reality prevails in evolution. But whoever has studied quantum theory sees that this argument does not go far enough. Statements about classical measurements, about established irreversible facts satisfy classical bivalent logic; from this it does not follow that this holds also for a description of the possibilities. Quantum theory does, of course, stay but theory. It corrects too narrow alternatives by more comprehensive alternatives and determinism by mathematized probability predictions. It exposes theory as an open approximative procedure.

Practice is, first of all, a residual category invented by theoreticians for the description of normality. The world is full of not-elephants, human life is full of not-theory; it is called practice. But then theory soon begins to study the structures of practice and thus to refashion and represent it on the binary scheme of the true and false. Aristotle very wisely delimits under the name practice that which bears its telos, its sense and purpose, within itself; the rest is called poiesis. The guiding value of practice is called the good. In the Christian culture, this is connected with the Jews' gift to humanity, the passionate distinction between good and evil. Morality, which rises as a summit on this base, is from the very outset primarily political morality. Economic and political power is a humanum, an—in principle—unlimited accumulation of means, first feasible to man, for the purpose of social enforcement. It is, seen from the standpoint of nature, a world-transforming cultural luxury. The pathos of political morality is, so to speak, a counter-luxury against luxuriating power. The enlightenment of European modernity rationalizes morality, lends it thereby an explicable claim to general validity and consequently powerful political clout. This is a victory of theory in the field of practice. Ambivalences are not lacking. The conviction of fighting for the good legitimizes many crimes. Self-righteous theory becomes untrue. Self-righteous morality becomes evil.

Art is a special summit in an area that is neither theory nor morality. Song, dance, and flute-playing, carving and architecture, story-telling and drama have

[4] Cf. Chap. 5 in this volume.

existed since time immemorial. That all of this is art is a smart discovery by theory. What art actually is, however, it seems to me, theory has never really been able to say. As a plank in the scaffolding I propose this homemade formula: Art is the blissful perception of form through the creation of form. For those who might be unwilling to use the adjective blissful, in view of the current earthquakes, I offer as an alternative 'seismographic'; in both cases, it is a matter of vibration. Judgmental theory and commanding morality are power-formed. In our culture today, the aimless perception, the unjudged affect flees into art. Art today is, seen socially, the private and therefore tolerated refuge from the world of the will and intellect. The point of it is, though, to see that it shows us truth which escapes theory and morality. I have made only one, philosophical comment on it. The great mathematician Gauss once wrote in a letter about the "unutterable satisfaction of scientific work". Mathematics is perception of form through the intellectual generation of form; if it is also blissful, then it is art. And philosophy formed under the paradigm of mathematics. Art, the luxury of form, is perhaps also a generic term over theory and morality.

We have not come face to face with the unity of truth. Thinking it was the claim of metaphysics. In the social reality of cultures another power has preserved unity: religion. If one delves into the details, one can—to use the current way of speaking—perhaps distinguish four social roles of religion: religion as bearer of a culture, as a theology, as a radical ethic, as an inner experience.

Religion as bearer of a culture is an expression in retrospect. While religion was bearing our culture, it was not functionally understood as a culture bearer but directly as truth, as the omnipresence of the divine. The three other roles, however, are attempts to bring out the crux of religion in a world of shattered tradition.

Theology is the determination of the true and false in religion, it is theory. Theology is, at first, the terminological name of the heart of metaphysics. Christian theology, thus I would venture to say, is the struggle between the Greek gift to humanity, the elucidation of true and false, and the Jewish gift, the revelation of good and evil. Elucidation and revelation—two metaphors of light.

Radical ethics is initially, in my present way of speaking, morality. With Plato and the Jewish prophets it is political morality. But an entirely different experience upholds it: the luxuriating, bliss-engendering self-denial of the ascetics. Nothing grand happens without a spiritual surplus. The incontrovertible truth of the Sermon on the Mount does not first manifest itself in its imperatives, but in the indicative of the Beatitudes.

We are thus referred to the inner experience. We are back in the third region not exhausted by judgment and action, by theory and morality. The inner experience of religion is belief, ritual, prayer, also ecstasy; it is the transformation of the person. This reminder about the ascetics leads us historically to a third metaphor of light, a third gift to humanity: Indian spiritual enlightenment about the unseverealness of opposites, experience of the one.

I started out from the Greek attempt to think of the one. Thus the circular path has closed once.

We are philosophizing today. I know of nobody who could say now how time, physics and metaphysics, or how elucidation, revelation or enlightenment fit together. Perhaps our next experience will not be contemplative but the self-bred crisis of political history, of power-formed culture.

Carl Friedrich von Weizsäcker 1992 at his "Alm" in the Alps. © Lili Bansa who granted permission to use this photo

Chapter 5
Biological Preliminaries to Logic

5.1 The Question

Is it not inconsistent to keep referring to the Platonic tradition of philosophy as relevant for this book on the one hand and, on the other hand, to pursue a biology of the subject?[1] The formal answer to this question is contained in our earlier discussion of the image of the circular movement. We now seek a stronger contextual statement of what we see in the circular movement.

The fundamental thought of Attic philosophy, of Aristotle as much as of Plato, is that only the *eidos* is recognizable. The *eidos* appears in later philosophy under the term 'concept'. A philosophy that maintains the priority of the concept stands opposite that tendency of the medieval-modern tradition that designates itself as nominalism in logical respect and as empiricism in epistemological respect. It holds the individual thing or the individual fact as the primary given and attempts to understand the concept as a designation of the particular. It would be pointless to want to decide the dispute of the argument between such vaguely characterized philosophical tendencies. In the attempt to formulate precisely the positions, to be able to discuss the issues at all, it would be necessary to invest so much philosophical work that it would make much more sense to spend the effort on introducing new concepts that no longer need the old names. But to clear the path to these new concepts we should attempt to understand what the old concepts have meant in their assumed clearest phase. Presumably, they were clearest at the beginning, when a single philosopher introduced them to designate phenomena he saw and when they had not yet reached the stage of historical multiplicity that

[1] This text was first published in: *The Ambivalence of Progress* (New York: Paragon House, 1988), pp. 181–194, and is a translation of: *Der Garten des Menschlichen* (Munich: Hanser, 1977): II.6.

M. Drieschner (ed.), *Carl Friedrich von Weizsäcker: Major Texts in Philosophy*, SpringerBriefs on Pioneers in Science and Practice 23,
DOI: 10.1007/978-3-319-03671-7_5,

necessarily accompanied their introduction into different philosophical systems. Because every concept establishes its meaning only in context.

The dispute about the priority of the concept or the particular usually takes the form of the categorical judgment or, as we commonly say today, of the predicative sentence. Therefore it takes on the form of logic. The sentence "S is P",—for example, "The lion is a predator" or "Socrates is a philosopher",—says about a subject (lion, Socrates) what its predicate (predator, philosopher) is. In the so-called singular judgment, the subject is a particular (Socrates). The predicate is, on the contrary, normally not a singular thing, but a concept. The question as to the relationship between the particular and the concept is therefore, in part, a question as to the meaning of logic. Why does the predicative sentence form even exist? What is its logically so fundamental meaning? Why do sentences, concepts, proper names exist at all?

To what authority should one turn to answer such questions? Tradition holds that logic is self-evident, that it is certain a priori. If that is so, the authority for our question would be this evidence, the insight that we have and owe to the certainty of logic. The search for evidence in support of authority normally leads no further than to a more precise repetition of the assertion, to an assurance that it is really so. The problem becomes more acute if we ask what kind of certainty really exists a priori in the case of logic. Kant has introduced the difference between analytic and synthetic judgments. He illustrates them even with the predicative sentence form. An analytic judgment is, according to him, a judgment whose predicate only states something that is already contained in the concept of the subject, while a synthetic judgment adds something to the concept of the subject. One also sees that Kant must think of the subject, too, as determined through a concept in order to be able to explain his differentiation. It appears obvious to him that analytic judgments are a priori certain, just as judgments a posteriori are synthetic. His problem is the synthetic judgments a priori. Let us leave this problem aside and ask now what constitutes the certainty of analytic judgments.

Modern logic retains the concept of analytic truth as a concept of reflection (a concept of the metalanguage in which we speak about logic). But it attempts to separate it from the connection to categorical judgment (and from the imprecision of 'a concept within a concept') and approximately defines an analytically true sentence as a sentence that is true through its mere form. If we ask, however, what sentences are true through their mere forms, we are referred to the rules of logic. If we then ask why the rules of logic are true, we are again referred to evidence. But we must assume that the rules of logic are not, in turn, true merely through their form. If they were certain a priori, then they, the rules of logic, themselves would be the simplest example of synthetic judgments a priori. But the synthetic a priori is highly suspicious to most philosophers of logic today. This short summary has perhaps been useful to make understandable the claim that the foundations of logic are philosophically not clarified (which I would be prepared to defend in a larger context).

Perhaps it is in this situation that one can attempt the method of circular movement. What is the origin of logic in life? From which historical situation does it come? And what is its biological background?

5.2 The Place of Logic

The question, "What should be the beginning of philosophy?"[2] is asked only after the beginning of philosophy has already been made *historically*. What, then, does this question mean? It concerns the *systematic* beginning of philosophy. It thus presupposes that philosophy should have a systematic beginning. It concludes that from a historically given knowledge. From which? I maintain: from the model of deductive sciences. But of these there exists, historically, so far only one, namely, *mathematics*.

Let us leave open the question of how much the transference of the structure of mathematics is worth to philosophy, and let us ask instead about the basis of this structure. Deductive mathematics appears as a system of axioms and theorems; the theorems are arrived at with the help of *logic* from each other, and finally from the axioms. The structure of deductive mathematics is therefore determined by logic. What is logic? It is the doctrine of *conclusions*, of *judgments*, of *concepts*. The order of these three words begins with what is first the conscious purpose of logic that of correct conclusions, and then proceeds to what as a component of correct conclusions becomes observable through continued reflection.

The next question must be, "How come there are concepts, and their combination in judgments and the combination of these in conclusions?" The 'metaphysical' answer is the doctrine of ideas. A 'positivist' answer, which first looks at the unquestionably given, finds concepts in the form of words, judgments in the form of sentences of a language, conclusions as parts of texts. Logic is then the system of rules about *language*, a chapter from *grammar*.

Linguistics today is a booming science. It discovers, in its way, the a priori in the form of those rules without which nothing meaningful could be said.[3] If logic formulates a part of these rules, through what, then, is it distinguished, if at all, within the system of grammar?

At this point in our reflection, the question must be raised as to what constitutes the *simplicity of logic*? When we asked about the beginnings of philosophy, it appeared natural to assume that what determines this beginning as beginning is at the same time the simplest. But now further reflections have led us into the immense field of linguistics. Between the immensity of sciences whose contents

[2] Hegel: *Wissenschaft der Logik*, [The Science of Logic [English] translated by George Di Giovanni (Cambridge: UP, 2010) First Book: "What must be the beginning of philosophy?"].

[3] I am indebted for essential information to Ernst Tugendhat: *Vorlesungen zur Einf□hrung in die sprach-analytische Philosophie* (Frankfurt: Suhrkamp, 1976).

were arrived at with the help of logic and the immensity of the grammatical structures that exist behind logic, logic itself appears as a 'bottleneck' of simplicity. From where does this simplicity come?

I assume that it comes from the reference of logic to *truth*. Every declarative sentence must state the truth. But whether or not it is true does not depend on its grammatical form. Logic, however, formulates requirements for the form of utterances so that they can be true at all.

A theory of logic thus requires a theory of truth.

5.3 Practical Interpretation of the Bivalence of Logic

As a pragmatic definition of truth, I will designate for the following its definition as the adaptation of the insight action to the circumstances. This definition does not make the claim of elucidating the nature of truth. It concerns one aspect of truth. It grasps logic as a theory about speech acts and speech acts as a specific type of activity. In its application in this chapter it observes of human activities essentially those traits that they have in common with animal behaviour. The thus defined truth refers then to nothing but the correctness of behaviour; everything that for human truth touches on the capacity for dialogue and reflection is ignored here. But we will see that certain characteristics of logic that are almost always accepted as given, such as the division into sentences and concepts and the bivalence of sentences, have their precise correspondents or precedents in the structure of animal behaviour. This correspondence is rooted in the cognitive aspects of life processes.[4] One can thus speak, in a strict sense, of the biological preliminaries of logic.

The basic fact of logic and of grammar is the sentence. It is a unity, often complex. As unity it is characterized, insofar as it is a declarative sentence,[5] by having, in its entirety, a 'truth value', being either true or false. Other sentences, such as commands and requests, have mostly a similar unity of intention. Through the sentence intention—meaning the intention of the speaker who means the word or words as a sentence—and, in cases of declarations, through the truth value—meaning the fact that what is said must be either true or false—one can also differentiate between a sentence that consists of a single word and the mere uttering of that word.

What, actually, makes the unity of the intention of a sentence? We should ask this mostly in regard to a declarative sentence. In the case of a theory that demands conformity with the truth, unity would mean unity of the intended matter of fact. But if one attempts to say what a matter of fact is, then one encounters difficulties that may only be overcome by reversing the explanation: Matters of fact are what

---

[4] "gnoseomorphic character" see for: "Erkenntnisförmigkeit" in Chap. 4 of this volume—*Editor* [MD].

[5] Λόγος αποφαντικός, Aristoteles: *De interpretatione*, 4, 17a2–3.

can be said in a sentence.[6] But with that the unity of the sentence intention becomes all the more a problem. Matters of fact that one could capture in a sentence appear continually to merge with each other, and only the understandable but never fully articulated sentence intention picks out from the continuum something that is usually not well delineated. For example, "It rains." Where? 'Here.' That means in the vicinity of the speaker. "All humans, by nature, desire knowledge" (first sentence of Aristotle's *Metaphysics*). What do they desire? What is knowledge? Do they always desire this? Do all really desire this? No, not really but by nature. A law is established to make the declaration true. Are the facts really verifiable or falsifiable? But from such sentences comes knowledge. Can one explain the unity of such matters of fact, as in these two examples, other than through the unity of sentence intention?

Another problem that relates to the unity of the sentence is the bivalence of logic. Every declarative sentence is either true or false. What is the matter of fact if the sentence is false? The matter of fact stated in the negation of this sentence. But are there really negative facts? "It does not rain" is the only possible answer to the question "Is it raining?" or, stated pedantically, "Is the matter of fact such that it is raining?" Bochenski says about the negation of predicates, "The world is full of non-elephants." Negation is a characteristic of logic required for explanatory purposes, not a fact of reality. Plato saw this clearly in his *Sophist.*

I propose to understand the declarative sentence fundamentally in the sense of the pragmatic definition of truth. A declarative statement is exactly true when a certain course of action would lead to success.[7] Precisely when it rains do we have need of an umbrella. Precisely if all humans desire knowledge by nature does it have meaning to write *Metaphysics.* Precisely if 2 times 2 equals 4 must one prepare four beds if two couples are coming to visit. Of course, these examples are simple stylizations, comparable to the uneducable primitiveness of innate behaviour patterns. The assertion is only that even our subtly testable declarative statements have content through the methods of their test. But the subtlety of a pragmatic epistemology can be developed only after we know to what extent the pragmatic definition of the truth explains fundamentally the unity of sentence intention and the bivalence of logic.

Here, we could go back to a characteristic of behaviour theory that is first given empirically. A behaviour pattern in animals is, in the simplest cases, a sequence of activities whose order is innate and whose discharge follows in accordance with external circumstances. There is, in the simplest examples, no intermediate stage between the occurrence of this discharge, with its resultant behavioural response, and the non-occurrence of both. If we look closer, an occurring intermediate stage can be the result of either a poorly functioning pattern or a complicated pattern in which numerous opportunities for decision exist, possibly under the influence of learned procedures. Since both of these refinements of the basic pattern do not

[6] This expression goes back to Strawson.

[7] All this is discussed with much more precision in the work of Tugendhat.

deny its structural validity, let us remain, for the sake of short expression, with the simplest example. Then the decision about the occurrence of the discharge in a certain time period is a simple yes-no decision. Also, it possesses the asymmetry that we find between being and nonbeing. If the discharge takes place, then the behavioural response occurs. If the discharge does not take place, "nothing" happens, and this requires an inquiry into the disappointed interest of the respective observer, to find out what it was that did not occur.

But the behaviour pattern, with its all-or-nothing structure, is not yet a sentence, neither imperative nor assertive. However, we can analyse from it the structure of sentences. We, the behavioural scientists, can define "a matter of fact for the animal" as those factors that have caused the discharge. Uexkull's *Umwelt* consists in such a matter of fact for the animal. Human thinking is, then, the re-presentation of possible activities with their consequences, be this in language or be it, for example, in visual fantasies. How this representation is possible is a highly complicated theme of a descriptive and causal theory of speaking and thinking. But it appears plausible that such represented activities have also the basic yes-no characteristic of elementary animal activities, though in a highly complex form, which yet permits a reduction to its simplest elements. A language representation for humans of such a represented "matter of fact for humans" would then be a declarative statement. Here the phrase "for humans" appears twice: A matter of fact for humans is represented in language for humans. The usual reflection of language logic recognizes the "for humans" of the language but not that of the matter of fact, and thus takes the unity of linguistically represented possible actions and its all-or-nothing structure as the unity and the being-or-nonbeing structure of the matter of fact.

In a simplification that brutally suppresses the many inherent complications of human speech and thought we can state the origins of the bivalence of logic as follows: The declarative statement represents a matter of fact. A matter of fact is defined through a suitable pattern of activity. The pattern of activity has an all-or-nothing principle: It can happen or not happen in a given situation, but not both at the same time (law of non-contradiction for activities) and not *neither* of both (law of the excluded middle for activities). The bivalence of the declarative sentence is, itself, an insight of reflection. In 'simple' statements we express the matter of fact that exists. A non-existent matter of fact is not expressed. But almost all existing circumstances, too, are not stated, namely, when they are without motivation for action or when no motivation exists to communicate their possible action motivations. Every simple statement is 'meant to be true'. But it is possible to doubt every simple statement, but not all at the same time. A doubted statement is reviewed through a process that can be expressed in a number of again simple statements. This process tests if the asserted matter of fact actually exists. The original simple statement (call it p) becomes now a reflected statement. If it proves to be true, then we can designate p as true, and we can also say "I maintain that p" or "p is true." Otherwise, "I maintain that not p" or "p is false."

This depiction, I hope, also explains a number of phenomenologically known characteristics of our perception in causal respect. Logic is bivalent because it refers to activities that can be either done or not done. If we call thinking in the context of logic 'intellect' and the conscious, decision-making active motivation 'will', then the correlation of both follows: Intellect can think what the will wants, and the will can want what intellect can think. The bivalence, the division of reality into alternatives, is not a property the world shows us without our assistance; it is the way we successfully grasp reality. Intellect is power-shaped. But the bivalence of logic is valid only for reflected statements. Through the grasp of doubt (seeing *two* possibilities: doubt–double) the isolated simple statement becomes a reflected statement. For simple statements and for what is not said, the predominance of truth over falsity is valid, what Heidegger phenomenologically describes with the concept of unconcealedness. A fringe of an indeterminate number of possible simple statements is indispensable for every single act of doubting reflection. This fringe results from the mostly non-explicit perception of what is given, and what can only in small fragments be dissolved into matters of fact. This is the predominance of perception over intellect. In perception, we are already informed of the whole. Therefore we are able to act in detail toward specific goals. Perception of the whole that disposes freely with conceptual thinking I would like to call reason. In this respect, reason is the precondition for activities of the intellect. Emotions are then something like reason before the development of the intellect. Therefore we talk about the rationality of emotions.

We end this section with another philosophical reflection about the meaning of our selected procedure. A matter of fact has proven to be "a matter of fact for humans". It is by no means the philosophical intention of the present notes to reduce the world to our ideas. The theory of animal behaviour offers a better model. Behavioural scientists know that in the "matter of fact for animals" there are essential aspects of the kind of reality that he knows as a natural scientist (and which he is convinced he knows better than the animal does). Critical reflection means that we test also our own respective and preliminary knowledge for its truth contents, its conditions, and its boundaries. A philosophy like the one attempted here reflects critically not only on particular knowledge but on the form of knowledge itself. It asks why knowledge is presented in statements, and even in negatable statements. It is methodically not illegitimate if it thereby refers back to empirical facts of the ancestors of homo sapiens (more precisely, of the present descendants of such ancestors). It is neither illegitimate that it uses that method of knowledge whose very origins it itself studies, so that it articulates its speculations in statements, in statements that are negatable. This is exactly what is meant in the 'philosophical genitive' which is simultaneously *genitivus subiectivus* and *genitivus obiectivus:* "The critique of pure reason" is a critique of reason through reason. Only our essentially historical way of asking forbids the division of the contents of our knowledge into empirical and transcendental; our transcendental reflection uses empirical material with the eventual goal of finding that its empirically discovered laws are transcendentally necessary.

In this sense it is, naturally, also legitimate to ask whether the depicted discharge scheme of behavioural events has itself a basis in things, such as, for example, a selective advantage in the world as it is (whereby we furthermore describe the world “as it is” in negatable statements). Let us use the preservation of the species as the criterion for success. Something, namely the species, and in the short term also the individual, should remain preserved. Therefore, the activities must maintain, as constant, certain properties of the one who acts with a certain unambiguity or must reproduce them after a while. This maintaining happens by itself as long as the conditions are unchanged. It happens also by itself as long as the conditions change only to an irrelevant degree. Therefore, there must be a mechanism that normally does not respond, but which does respond to relevant changes of the circumstances (external or internal, for example, enemies or hunger). Finite beings such as the organisms can have only a finite number, in fact, only a small number, of reliably functioning, relatively complex active and dischargeable patterns of behaviour. This plausible statement can be quantitatively substantiated through the thermodynamic conditions of reliability, namely, adequately certain irreversibility. Here we run up against the same arguments as when dealing with necessarily irreversible processes in measuring instruments according to the quantum theory of measurements. Just as classical physics presents itself to us through measuring instruments, here the unity and negatability of circumstances, thus classical logic, presents itself to us through the functional conditions of the physical object, which is expected to act in a manner that is structurally like acts of understanding.

With this reflection we have, of course, not escaped from the historical circle of our own knowledge. The finiteness of the reaction methods of an organism is the physical result of the fact that it consists of a finite number of atoms and that these have only a finite number of reproducible modes of reaction. The law describing the facts of this situation is, for us, the quantum theory. We try to derive these laws from finitary axioms; and their finitism expresses the finiteness of a thinking that is dependent on uniform sentence content. By attempting to express this in negatable propositions, we also think of the possibility of the negation of these propositions, thus of nonfinite matters of fact or of a reality that is not reducible to ‘matters of fact’. We can grasp the thought of infinity only as potentiality, i.e. as future. In a certain sense, modal logic for the future, such as quantum logic, is not bivalent. As a contribution to the necessary discussion of these questions we wanted here to suggest a probable reason for the bivalence of classical logic.

5.4 The Basis of the Subject-Predicate Structure of Sentences

That the elementary declarative statement predicates something of a subject (ὑποκείμενον), namely, either giving or denying a predicate (κατάφ ασις–ἀποφασις), has been taught since Aristotle, in close connection with the unity of sentence intention and with bivalence. Plato, in *Sophist,* already presupposed this structure of the sentence. But the relationship of both structures is by no means evident. We also recognize sentences as true or false that are constructed in a complicated manner from elementary sentences, as well as sentences whose simplicity does not suggest an immediate meaning in the sense of the *S-P* structure, such as "It is raining," 'Fire!,' "It is not raining." This means that the unity of sentence intention, bivalence, and negatability do not result in the *S-P* structure; the latter is a limitation of the multiplicity of conceivable sentence structures. On the other hand, this structure is very widely used; and if I understand Chomsky correctly, he sees it as an indispensable element of deep structure even where it does not appear in the surface structure. Is it possible to comprehend such universality?

In the Aristotelian syllogistics, *eide* are built upon *eide*. Tied to this is the classical notion of the generic term, of the pyramid of notions, etc. According to Detel, Plato already constructed predicate sentences. But among Snell's three examples of predicate sentences

The lion is a predator,
The lion is yellow,
The lion roars

this model applies only to the first. Properties and activities are only forcibly made into generic terms. Essential to them is that they do not always or essentially or necessarily apply to the subject. "The lion roars if he is not silent",[8] and a lion turned gray is a lion, too. The limitation of the predicate structure to the timeless relationships among *eide* does not only exclude the fullness of the time-related language, it also blocks the way to temporal logic. We suspect today that the *eidos* structure is, on the contrary, a special variant of the predicate structure, that, for example, *eide* are predicates that have become timeless.

We find a temporal foundation of the predicate structure in Aristotle himself, in his *Physics* (*A7*). Aristotle shows that if there are changes there must exist a representation through two or three principles, which he calls matter, form, and privation (ὕλη, μορφή, στέρησις). The bronze (matter), which is not a statue (privation), becomes a bronze statue (form, namely form in matter, ἔνυλον εἶδος). Abstractly stated: Whoever wants to describe change must indicate something that remains unchanged (if nothing remained the same, then a conceptual description would be impossible) and something that changes. What remains the same is called matter, and what changes is its form. As form changes, it assumes two appearances: *A* becomes *B*. The usual description designates one of the two appearances as the intended form and designates the other only through its absence. In this regard, the two-principle structure of matter and form can also be understood as three-principle structure, of matter, form, and privation. One sees here directly the relation aspect of the concept pair matter-form. In relation to a certain change, bronze is matter, namely, in relation to what does not change. In relation to those things from which it is made, copper and tin, it is form. Matter

[8] "Der Löwe brüllt, wenn er nicht schweigt." Wilhelm Busch: *Naturgeschichtliches Alphabet für größere Kinder und solche, die es werden wollen.* In: Rolf Hochhut (ed.) *Sðmtliche Werke I*, pp. 111–122.

that is the subject of a declarative sentence is itself an *eidos* because only an *eidos* is expressible. *Prima materia*, πρώτη ὕλη, is an abstract philosophical concept, not an actual thing.

It belongs essentially to the Aristotelian concept of matter that matter means possibility (δύναμις), namely, receptive possibility (δύναμις τοῦ πάσχειν). Bronze can be made into a statue, but it need not. The change is not described as a determined course but as the realization of a possibility, which need not have happened. This implies, by the way, neither positive nor negative prejudgment of determinism. It is conceivable that whoever knows all interacting causes could also know that this bronze must become a statue. But that is not the actual situation of man seeking knowledge. One needs composable concepts, thus possibility: matter as possible realization of a form, form as possible form of a matter.

In a temporal logic, it makes sense to use an analogous scheme. We speak of decidable alternatives. Here we must first indicate the alternatives, that is, differentiate them from other alternatives, and, second, indicate what the decision is. The first corresponds to matter, to the possibility of that form or its privation, and the second step corresponds to form itself. A form is only identifiable if it appears often (confirmation). A possibility is only identifiable as possibility of a form. But its being a possibility means that the form may exist or not exist in the present case (unprecedentedness).[9]

Under this concept of predication fall the three examples of Snell, once the concept of a constant subject (οὐσία in Aristotle's sense, πρώτη οὐσί of his 'Categories') has been justified. This justification, however, needs further development. Subjectless sentences such as "It rains" indicate that the phenomena do not have to show permanent subjects along with applicable concepts. Methodically speaking, the subject is a permanent alternative (the possibility itself as form: for example, the piece of bronze). But there is, we assume, no transcendental proof that permanent individual objects exist. This may be seen through a detour to physics. Elementary particles are the simplest quasi-individual objects, but even they are no longer the original ('Ur'-)alternatives.

Snell's three examples appear to lie much closer to a world of individual objects. An object can have properties that either can or cannot change. If they cannot, one calls them intrinsic characteristics and groups them under a substantially expressed generic term. If they *can* change, one designates them with adjectives. Change of their properties occurs usually through outside influences. The causal concept of influence indicates the determining character of temporal processes in a world of individual objects: One works upon the other. To be the cause of such an influence is called action and is described in Snell's third example.

---

[9] Cf. Ernst Ulrich von Weizsäcker: "Erstmaligkeit und Bestätigung als Komponenten pragmatischer Information", [Unprecedentedness and Confirmation as Components of Pragmatic Information], in: E. v. Weizsäcker (Ed.): *Offene Systeme I* (Stuttgart: Klett, 1974).

That we describe the timeless relationships through predicate sentences is the least obvious. Negation is natural in temporal circumstances, where it expresses real possibilities. The negation of a timeless true sentence ("two times two is not four"), as a real possibility, expresses only the possibility of ignorance (of error, of lie). We can disregard here this central question of philosophy.

In modern philosophy of logic, in view of the background of Aristotle's substance ontology, and strengthened through the school of thought of nominalism, one has chosen a construction entirely different from the one attempted here. One has presupposed or attempted to prove as immediately obvious what is in need of justification, or what has only limited truth: the existence of isolatable objects of permanent identity. Objects are designated by singular terms, by proper names. The philosophical problem then begins with the introduction of concepts, that is, predicates. The name 'Socrates' designates that Athenian philosopher who was Plato's teacher and who drank hemlock. But to what do predicates refer such as 'philosopher', 'Greek', and 'mortal'? We say they refer to a class of objects. But what is a class? Is it a totality of objects? One view in a narrow nominalist sense would be that a predicate is only a common name for all objects of a class. Frege and Russell have pointed out the difficulties that arise if one does not distinguish a class from its objects. It can be said, in the scope of a temporal logic still to be constructed, that we can never know completely all the empirical objects that fall within a class. At least we cannot know the future ones. But we can already know the predicate that determines whether they belong to the class when we will know them. Accordingly, the concept of the predicate takes precedent over the concept of class.

Frege's and Russell's thesis, that predicates may occur as subjects of predicates of higher levels and thus form something like objects of higher levels, is sometimes designated as 'Platonism', especially in English-language philosophy. But this is a historically and systemically absurd designation. This so-called Platonism accepts unquestionably the existence of elementary objects—something Plato consistently denied. It then conceives of concepts according the example of these objects—the second mistake that Plato tirelessly fought against. In the Aristotelian syllogism, there are no examples of singular *termini,* only of quantified predicates. It is not "Socrates is mortal" but "all Greeks are mortal".[10] In order to approach the meaning of Plato, one must first understand every singular *terminus* as a characteristic, that is, as a predicate under which falls precisely one object. In this sense I have explained above the meaning of the name 'Socrates'. Only then can one, like Kant in his explanation of analytic judgments, talk in the traditional way of the respective concept of the subject of a predicate sentence.

Contemporary analytic linguistic philosophy sees itself pressed step by step into the same direction. I paraphrase here from Tugendhat's presentation. If we search for the meaning of a sentence in the conditions of its use, we can first explain

[10] Cf. J. Lukasiewicz. Aristotle's Syllogistic from the Standpoint of Modern Formal Logic. Oxford: Clarendon Press, 1951. Also: G. Patzig. Die aristotelische Syllogistik. Göttingen: 1959.

sentences that are connected to a situation and that attribute a predicate to the present situation. Then the introduction of singular *termini* takes place to make the sentence independent of the situation. The designation of objects is connected by Tugendhat to the spatial-temporal situation, or, as the quantum physicist must say: to the concepts of classical physics.

Keeping the logical problems in perspective, we return to the biological analysis of the perception of concept and individual case. We can say that *for us* a concept corresponds to the simple behavioural pattern of animals. *For the animal* the respective behaviour may be hardly more than the undifferentiated unity of the three moments that we separate into concept, emotion, and action. We have such unbroken behaviour, I believe, only where we don't even notice our own action. It is only where a higher degree of complication of behavioural patterns can restrict or permit a definite course of action in accordance with a higher criterion, that there can be a biological need to view this process *as a possible* process. It probably is here that the separation of action and emotion begins. Emotion, in a way, is the perception of not completed activity as being indicated but not yet completed. Here perception separates from motion. The concept as concept, thus the separation of concept and individual case, of thinking and perception, is an accomplishment that higher animals occasionally achieve, but which is truly achieved only by humans, articulated by language.

Here are the roots to an understanding of the philosophy—above all Platonic philosophy—of a fundamental phenomenon that I would like to call *co-perception.* It is the co-perception of the concept in what falls under the concept, which is, in the simplest example, the individual case. "There goes the ball!" Have I perceived the individual thing that is included within the concept ball? Or have I perceived the *eidos* (the concept) ball in one of its manifestations? This manner of inquiry is wrong. I have perceived *this as ball.* If *we* speak of the perception of animals, we must say that they perceive concepts. If the young bird in the nest opens its beak when it notices the approach of the mother with food, but also responds the same way to the approach of a cardboard dummy: Does it confuse the dummy with the mother? Presumably not, but it reacts to the concept of the "feeding beak." It is the accomplishment of philosophy to distinguish the co-perceived concept from the individual case. Philosophy is a training to perceive both in its own characteristic way. The perception of the individual appears as perception in the narrow sense since the time of Greek philosophy and continues, more narrowly, in the nominalist-empirical tradition. In Plato and Aristotle, the thinking of a concept is still designated with a word that refers to optical perception (νόησις). The problem of empiricism, how special experiences can be the basis for universal concepts and universal judgments, is solved in Aristotle's theory of induction (ἐπαγωγή), in which he says that we perceive the universal *in* the individual case. The gathering of isolated data can be helpful but, in principle, *one* individual case is enough for induction. Here the phenomenon of co-perception is still directly described from the untroubled phenomenological view of Greek philosophy.

The task of philosophy can be depicted, stylized a bit playfully, in the following steps. If the individual case and the concept are differentiated by the reflecting perception, then in the individual case of this perception the difference between the concept of the individual case and the concept of the concept is co-perceived. The *eidos* is, described from the individual case, the regularity that we co-perceive in the concept—in the possibility of the mode of behaviour that constitutes the respective concept. The co-perception of the *eidos* in the sensory object is the appearance of the *eidos,* in the sense in which Greek philosophy understands the concept of appearance (φαινόμενον). The co-perception of the difference between *eidos* and sensory object is the appearance of this difference. Philosophy, since Plato, has been the attempt to directly perceive this co-perceived difference. It is a meditation on this difference. Therefore, philosophy is theory of appearance: It is an appearance of appearance. Here one must co-perceive what is co-perceived in *eidos* that is perceived *as eidos,* and therefore is co-perceived: the essential elements of *eidos.* Thus, expressed in the Platonic tradition, the transcendentals: the One, the good, being, and truth.

The ascent from the cave is the path of this meditation. But it must be said again that only Plato's *descent* back is constructive philosophy. It is the reestablishment of the unity of *eidos* and individual thing, disrupted for pedagogical reasons at the beginning of the ascent. It is thus the recognition of all individual things as what can only be recognized: as *eide.* Because of the escapistic tradition of neo-Platonism only Plato's philosophy of ascent was welcomed, and this resulted in division, since the high Middle Ages, of the two movements of ascent and descent into two philosophies, the Platonic and the Aristotelian. This division is reflected in Raphael's *The School of Athens,* which apparently motivated Goethe to give a beautiful depiction of both philosophies in the historical part of his colour theory.[11] But the separation of the schools, in the radicalization of positions, leads to unsolvable problems. Empiricism holds that the data received from sensual impressions are given as such, and it cannot reconstruct laws from them; it does not understand that sensual data, because of our biological endowment, are bound to be given under a co-perceived concept. Idealism—which is, actually, an ambiguous term—maintains that the direct perception of the *eidos,* the accomplishment of a meditative abstraction, is an independent power that can only be illustrated through experience but not taught by it; idealism's historical destiny is to dogmatize the respective steps of empirically based insight. Idealism, so far, lacks the perception of history, the co-perception of time.

Reference

von Goethe, J. W., 1960: *Materialien zur Geschichte der Farbenlehre.* Hamburger Ausgabe, XIV (Hamburg: Christian Wegner): 53–54

[11] See von Goethe (1960).

Carl Friedrich von Weizsäcker (1987) in discussion with the editor [MD]. © Michael Drieschner who granted permission to use this photo

Chapter 6
Models of Health and Illness, Good and Evil, Truth and Falseness

This essay is the core of our reflections on cybernetics.[1] It is the only essay in this Part that tries to describe concretely what cybernetics can achieve. Written in the fall of 1967 but unpublished until now, it was inspired by a lecture on "Peacelessness as Mental Illness" delivered in Bethel in 1967 (published in *Der ungesicherte Friede*; Göttingen (Vandenhoeck & Ruprecht) 1969). On that occasion the question arose whether it is medically justifiable, or merely metaphoric, to call the inability to maintain peace an illness. As far as I can see, contemporary medicine lacks a sufficiently clear concept of illness to decide this question. Since I had no desire to merely skirt the issue, I was faced with the task of outlining a scientifically justifiable definition of illness. The result is presented in Sect. 6.2 of this essay; its social application is very briefly sketched in Sect. 6.4.

The formulation of the question is aimed from the first at the concept of truth (Sect. 6.5). This concept is presupposed in all theories concerning experience, including the theory of the unity of physics. If man is part of nature, then truth must exist within nature. Truth is unsymmetrically related to falseness. This relation may be explained by a comparison with the relation between health and illness. A further comparison that suggests itself is with the Platonic-Aristotelian philosophy of the *eidos*. It seems to me in character with this philosophy that cybernetic ways of thinking can be compared with it more easily than with any modern philosophy of consciousness since Descartes. I use this philosophy in interpreting cybernetics in this essay; conversely, in the essay "Parmenides and the Greylag Goose" (Chap. 7 in this volume) cybernetics serves in the interpretation of a Platonic text. Equally important as this relationship—once it has been recognized—is the contrast between a philosophy of the eternal *eidos* and a philosophy of irreversible time such as one must formulate in connection with the conception of 'progress' (Sect. 6.3).

---

[1] This text was first published in: *The Unity of Nature* (New York: Farrar Straus Giroux, 1980) and translated by Francis J. Zucker from: *Die Einheit der Natur*, (Munich: Hanser, 1971): III.4.

M. Drieschner (ed.), *Carl Friedrich von Weizsäcker: Major Texts in Philosophy*, SpringerBriefs on Pioneers in Science and Practice 23,
DOI: 10.1007/978-3-319-03671-7_6,

This essay, let it be emphasized, merely touches on the real problem; i.e., on the degree in which human truth differs from the being-in-truth of animals. Here we stand in need of a cybernetics of linguistic consciousness, of conceptualization, of reflection, of the Self. All this seems feasible, but so far most contemporary cyberneticists have been slow in taking up this line of research.

The following reflections regard man, methodologically, as a living being, the living being as a control system with feedback, and this system as the outcome of mutation and selection. In so doing, they attempt a biological anthropology, a cybernetic biology, and a Darwinist cybernetics. The reflections are designated as models since they cannot claim to state the truth about man, about life, and about the origin of the control systems. They are based on the supposition that such models can exhibit what can be objectified about man, about life, and about history. What the term 'objectify' may mean, these reflections do not ask. They simply try to present some examples of what objectification can accomplish, in order to provide material that might facilitate asking the question of what it is to objectify—and, in particular, to what extent objectification reveals truth, and to what extent it disguises truth.

The reflections also do not define the basic terms of the objectifying sciences on which they are based—i.e., general biology, cybernetics, and the theory of selection. They presuppose that the reader is somewhat familiar with these terms; this allows me to sketch rapidly, in broad outline, what my reflections might contribute to an anthropology that seriously regards the healthy, the good, and the true, as well as their opposites, as phenomena, and that tries to link them with objectifying science. To fill in this outline would demand an immense amount of specialized research, in the course of which many basic phenomena here omitted or unrecognized would surely show up. As far as these sciences are concerned, I am therefore merely sketching a strategy.

6.1 Health

So long as we are healthy, we do not notice that we are healthy. Health is one of those phenomena that we do not perceive as anything special because they are a part of everyday life. We discover them by their absence. Only illness enables us to see health as health. That I am healthy, and in what way I am healthy, is something I experience when, and to the extent that, I am or was not healthy, or when I perceive (but this it is difficult to do) that, and to what extent, someone else is or was not healthy.

Being healthy is not the only 'being' that escapes our gaze precisely by being, and that appears to us by its absence. One could say the same of being as such. In the present reflections we will encounter 'being good' and 'being true' as modes of being of exactly this sort. But the sought-for model of the phenomenon can probably be developed most easily by taking being healthy as an example.

The first time we recognize health as that which is missing in illness, we 'somehow' know what the term 'healthy' means, but without being able to define it. A healthy eye or leg is a well-functioning organ—an eye that can see, a leg that can walk. A sick eye functions improperly: it is farsighted or near-sighted or astigmatic or occluded—or it functions not at all, is blind. Health therefore seems to be 'competent functioning'. To define health, it appears necessary to know what 'function' means. Every organic functioning serves a larger whole: the whole man or animal or even group, family, or species. But what do we mean by 'serve'? How does an organ serve? The eye appears to have a purpose, it is there for man. Does man exist for something? Does the regress of purposes terminate in an end in itself, the regress of functions in a closed functional whole? Of man, too, we say that he is healthy or sick. Is a human being's health also a competent functioning? Where lies the purpose of his functioning? In society, perhaps? Can one speak of the health and illness of a society? A morbid age, a healthy nation, the schizophrenia of modern consciousness—are these metaphors, or do the concepts of health and illness extend that far? Medicine tends to equate health with normality. But what is the norm? Who sets it?

We study the replies to these questions offered by the cybernetic-Darwinist model of the process of life. I claim that the model gives precise answers, sometimes by deciding a question, sometimes by rejecting one as meaningless, sometimes also by making it clear why we do not (as yet) have the means of answering.

The cybernetic model of an organism views it as a control system; i.e., as an integrated system of control cycles with feedback. In each control cycle of a technical piece of equipment, man adjusts the quantity to be controlled to some desired (or 'set') value; the control mechanism then sees to it that the actual value never deviates from the set value by more than a certain permissible difference (the 'tolerance'). A control system, then, presupposes an integrated system of set values. At first we ignore the complications of the concept 'system' by in turn characterizing it in terms of a suitably abstract (say, multi-dimensional) set value, the norm around which the actual value fluctuates.

How is the norm determined in a cybernetic model that represents a living being? Evidently this question corresponds to the above question as to the purpose of a function or of a whole organism. Darwinism answers: The norm is the optimal value of the quantity in question, optimal in the sense of survival in the struggle for existence. The optimally adjusted individual or species (the one whose norm is optimally set) is the one that survives among competing individuals and species; the ability to survive is what defines which norm value is optimal.

To proceed cautiously, let us articulate the explanatory claim of the Darwinist hypothesis in three stages. All three claims are in fact made; we can distinguish between them conceptually:

1. The competition among individuals, much like that among species, plays an important eliminating role in the processes of life.
2. This competition suffices to explain the existence of a species-specific norm in the control systems of organisms, and to fix the value of this norm.

3. Mutation and selection even explain the origin of novel types of living beings; thus they explain causality in evolution.

Nobody will be able to deny claim 1, even if he rejects the cybernetic theory of organisms or wishes to define the norms other than Darwinistically. Claim 2 I will presuppose in the present section. That is, I do not examine whether it is correct but assume it as a hypothesis, and examine its consequences with respect to our original question concerning the meaning of health. Claim 3 will not be taken up until Sect. 6.3.

I now claim: Normality, which enables us to define health, is precisely what we have just designated as the set value, the norm of the control system that is the organism. What does this mean?

To be normal means to conform to a norm. This norm is customarily understood not as a private norm for a single individual, but as a common norm for a class of individuals. For the time being we ignore the differences among human beings. We speak, for example, of the norm that determines a biological species, such as the norm of a greylag goose.[2] The norm is what the zoologist describes in talking of 'the' greylag goose. Konrad Lorenz shows persuasively that no empirical greylag corresponds to this norm exactly, nor does the norm represent the statistical average of empirical greylags. The greylag of the zoologist is, rather, the Darwinistically interpreted Platonic idea of the greylag. It characterizes the bodily build and behavioural pattern from which an actual greylag must not deviate too far if it is to survive, and if it is to produce offspring capable of surviving. The norm of the greylag is the above norm of its characteristics as defined by the condition of optimal survival value. For the sake of simplicity, let us assume for the moment that such a norm is uniquely determined—i.e., we ignore the possibility of equally good variants, not to mention the possibility of a favourable further development.

What determines this norm in detail? Its survival value in the given environment, thus its adaptation, as one says, to the environment. The norm of a living being is not something that exists in itself; it refers to a particular environment, and within this environment to a particular 'ecological niche'. If one imagines all possible characteristics of an organism entered in a multi-dimensional parameter space, and represents the 'norm of a possible organism' by such a 'parameter vector', then the concept of the ecological niche implies the existence of certain values of the parameter vector that are optimal, in terms of survival value, with respect to all (or almost all) neighbouring values—for example, in an environment containing ants, a parameter vector for competence in ant eating. As shown by this example, the ecological niche depends on whether and how other ecological niches are occupied; ant eating offers an advantage only if there are ants. The norms of the various organisms are therefore not independent of one another. If we hold all external conditions—all the other norms and factors affecting occupancy of the pertinent niches—constant, we obtain what is called a constant environment. The

[2] Cf footnote 1 in: "Parmenides and the Graylag goose" (Chap. 7 in this volume).

health of an individual is then the optimal adaptation to the norm of life existing in the constant-environment niche that defines the species to which the individual belongs.

The biological term 'species' is the Latin translation of the Greek eidos, which derives from Platonic-Aristotelian philosophy. The concept of the norm just introduced is in fact a cybernetic model of the Platonic idea, and specifically so in the form Plato gave it, not in the form of the Aristotelian eidos. The norm of a greylag is not itself an empirically existing greylag. Nor does this norm exist in empirical greylags as their 'form'. At any rate, this is true only as a rough approximation, and it is really a metaphor. One actually recognizes an empirical greylag as a greylag by what it has in common with the norm of the greylag; i.e., by its participation (methexis) in the norm of the greylag. The norm therefore exists beyond (choris) all empirical greylags, but for that very reason it is what can be known, in the sense of the science of zoology, about empirical greylags. (Deviations from the norm can be described scientifically only insofar as they are instances of other general terms; i.e., of something like norms again.) The norm is not, however, a merely subjective thought that some people have. It is, first of all, not merely the name with which we subsume certain individuals under a class. The class of all greylags is not empirically countable. Rather, we recognize that an individual belongs to this class by comparing the individual to the norm. Nor is the norm a 'representation' that has been 'thought up'; viewed as a representation, the norm refers instead to an objective state of affairs: it is a 'true representation'. In fact when we speak of a norm, we mean the state of affairs it expresses, not our representation of it. This state of affairs is not an individual thing (not a goose), it is law-like in nature (the preconditions of the life of a goose).

As a historical aside, let me remark that this scheme probably comes as close to what Plato meant by the 'ideas of individual things' as a cybernetic model can come.[3] Plato's own examples of ideas are usually the highest concepts, such as 'the Same', 'the Just', or mathematical concepts such as 'the circle'. Only in a few instances does he mention the ideas of individual things—for example, in Book Ten of The Republic, 'the couch' and 'the bridle'. On comparing these last two examples, which are clearly structured alike, one finds that for Plato the idea of the couch was fashioned by God, and the idea of the bridle is what the rider understands, while the many empirical couches and bridles are manufactured by artisans (carpenter and harness maker, respectively) in accordance with the model furnished by the ideas. The parallel structure of the two examples implies that the rider knows what God has fashioned. But what is it the rider knows in this world fashioned by God? He knows the functional relation between horse and man in the activity of riding, which enables man to control the horse by means of a bridle. The rider knows, so to speak, the ecological niche for the invention of the bridle; from this niche stems the norm of the true ('healthy') bridle which the harness maker

[3] Cf. "Parmenides and the Graylag goose" (Chap. 7 in this volume).

must keep in mind.[4] Plato hints at the connection between the highest ideas and the norm of individual things only in the Timaeus, among all his published works; presumably he had worked out a more detailed hypothetical doctrine on this topic.[5] This 'derivational system' proceeds via the basic mathematical forms and the physical elements defined by them (the Platonic solids as quasi-atoms). The reduction of biology to physics that our era is witnessing—i.e., cybernetics—conforms with this scheme, as does the reduction of physics to simplest laws (Heisenberg's 'symmetries', in conscious analogy to Plato). What is hard to puzzle out in the Platonic tradition, and what has not been thought through in modern physics, is the connection between this mathematized science and what in our tradition is called mind and soul. In the context of our present deliberations, this problem lies concealed in the meaning of the concept of a model; we shall return to this near the end of our discussion.

6.2 Illness

Our definition of health has not yet clarified the notion of illness. The so familiar phenomenon of illness—influenza, cancer, and endogenous depression, for example—is something entirely different from the usual deviation of the individual from the norm. If health were defined as coincidence with the norm, and illness as any deviation whatever, then no healthy individual would exist. This would contradict the Darwinist sense of the concept of a norm, which in fact demands for empirical organisms a margin of tolerance about the norm; if a parameter can be measured quantitatively (such as body height, say), the survival value of the parameter in the vicinity of the optimal value will probably vary only quadratically with the deviation from the optimum. In short, the usual empirical distribution about the healthy norm is not what is meant by illness.

On the contrary, illnesses seem to have something like a norm of their own. Were it not so, the science of medicine, which classifies illnesses and recognizes them, which describes the typical course of illnesses, and, basing itself on experience, applies specific remedies, could not exist. In the Romantic age medicine taught that illness is itself an organism which comes into being, grows, and passes away in accordance with law. Bacteriology seemed to have revealed the kernel of truth in this doctrine: bacteria (and, with a grain of salt, virus) are indeed themselves organisms. But illnesses not explained in terms of germs develop no less lawfully. It even seems that well-defined complexes of symptoms (e.g., those of cancer) can have very different causes (chemical or radiological irritation, and still unknown 'endogenous' causes). It is not the cause that is specific in this instance, but the reaction of the organism as it appears in the pattern and development of the symptoms.

[4] I owe this interpretation to a remark made years ago by Georg Picht.

[5] Cf. Gaiser (1963).

If there is such a thing as a norm of illness, then we can ask what it has in common with the norm by which we defined health, and we can ask also how they differ. The norm of health defines the operating characteristics of a control system in accordance with a desired value set by the requirement of self-survival. Illness appears chiefly as a disturbance in the operating characteristics, as an operational dysfunction. But we do not call every such malfunctioning an illness. The malfunctioning must, to put it quantitatively, stick to a mean value. If it is too slight, it will not be conspicuous as an illness; it will be drowned out by the statistical spread, by the 'noise'. If the malfunctioning is excessive, it will bring systems operations to a standstill; that would be not illness but death. Furthermore, the mean value must be maintained throughout some time interval. To count as illness, the malfunctioning must neither return immediately to the healthy norm nor end immediately in death. In other words, the illness must maintain itself for a while as just this illness. It seems, therefore, that illness itself presupposes something like a control system that keeps it stable. Illness appears like a parasitic control system within the larger control system that we call the organism.

From the cybernetic point of view we may expect just such parasitic control systems as a result of the moderate perturbation of a highly complicated control system. Every sufficiently complicated control system shows a variety of responses to the steadily increasing enforced deviation of its actual from its set value. In the vicinity of the set value the system, if it is well-functioning, steers back to the set value; it will produce (usually damped) oscillations, or an asymptotic return to the set value. We can call this the tolerance span of health; the set value is the point of stable equilibrium within that range. The boundary of this range will be a point of unstable equilibrium. If a value exists beyond this boundary that represents another point of equilibrium, then the control system can, as the result of a sufficient deviation, start oscillating about this 'false', undesired value, or it can stay put there. Just as the 'true' set value was characterized by its survival advantage for the organism as a whole, so the false one will damage or weaken the organism. Thus the false set value is an illness that at first stabilizes itself; in the simplest case, it is the norm of an illness. In a chain reaction, the actual values of other control magnitudes are also perturbed. The illness, in other words, "takes its course", a course that ends in the return of the whole organism to the healthy norm, in death, or in a self-stabilizing sick norm—for example, in a 'crippling' or 'chronic', disease, or at least in a 'scar'.

In summary, non-lethal perturbations of a complicated control system are bound to show up as parasitic control systems. To say this another way: a highly ordered system can react to perturbations only in an orderly manner, assuming that it can still react at all. Downright disorder cannot occur. In this sense, the possibility of illness is a necessary consequence of the possibility of health. Illness could thus be defined as false health. Viewed Darwinistically, the concept 'false' designates the diminished survival ability; i.e., the loss in adaptability. Speaking Platonically: in a world shaped by ideas, even the Bad can come into being only in accordance with an idea.

6.3 Progress

Until now we have judged the control mechanisms merely by the standard of self-preservation. We now come to the third claim of Darwinism: that it can also explain the origin of the control systems we call organisms. This claim can hardly be proven empirically, and it can hardly be empirically refuted. In contemporary biology it serves as a heuristic principle. Again we do not discuss its justification, but assume it as hypothetically true and examine its consequences.

According to this conception, the totality of all organisms cannot, strictly speaking, be stable, nor can even a species be absolutely stable. A steady 'selection pressure' is maintained in the direction of the development of species that are more successful in the struggle for survival; besides, the evolution of some species results in a change in the ecological niches of other species. And indeed, in the case of highly complicated control systems one will hardly expect stable parameter combinations to exist.[6] We will omit concepts essential in the quantitative theory of selection, such as the finite step of mutation, the storage of recessive characteristics, isolation, and changes in environment. It is certain that the mere possibility of progress defined by improved chances of survival renders all norms oriented merely toward preservation systematically unsharp. As a result of changed environmental conditions, or of the coincidence with another deviation from the norm, a 'norm of illness' can suddenly turn into a 'norm of progress'. The human foot would mean illness to an arboreal monkey; for the ape treading the ground it is the road to an upright gait with all its advantages. This is how progress relativizes, to a certain extent, the difference between health and illness.

In the case of a being like man, who is capable of storing experience, the acceleration of progress is revolutionary. ("Man invented the inheritance of acquired characteristics.") As a consequence, a revolutionary relativization affects the distinction between health and illness, although the distinction does not disappear. Let us first keep to the biological domain and ask whether there is such a thing as a norm of progress.

Here we encounter a basic difficulty. In the usual sense, the set value of a control system is by definition 'conservative'. The norm is that to which the control system is to return again and again. A 'new norm' is the revolutionary introduction of a new conservativism—the transition of one stable species into another stable species, for instance. To a certain extent, this can still be thought of in cybernetic terms. To be sure, already at this point ambiguities are encountered. A species can, for example, split into several different species that specialize in different ways, as the Darwin finches did on the Galapagos Islands. To use a metaphor: from a saddle or peak, roads can lead into many different valleys. The uniqueness of the norm no longer holds here; nor is it to be demanded on 'ethical'

---

6 An optimum probably exists for a finite number of parameters; but the number of possible parameters in organic life is at least 'practically' infinite; i.e. the past history of life has surely been too short to reach such a possible optimum.

grounds (in the sense of an ethics of good will toward the particular species), for precisely this specializing multiplicity of forms may be advantageous.

It is even more difficult to define a norm of progress in the face of unknown 'future norms'. Still, even in this case at least one possibly relevant observation can be made. If we classify the possible properties of organisms that offer an advantage in selection, we can divide them into three groups; they serve:

(a) the survival of an individual,
(b) the survival of a species, and
(c) the modification of the species.

We have so far considered only (a) and (b). The question concerning a norm of progress can be interpreted as the question concerning the properties that advance the modification of the species in the direction of higher viability.

One is of course bound to ask whether anything of this sort can arise in the blind play of mutations. An increase in the rate of mutations can also be harmful. Most of the existing species seem to be stabilized in their ecological niches. But at least in the case of environmental changes it may be desirable to quickly try out many mutations. A species that can do this has a selection advantage over the 'slower try-out' species. I would like to mention at least one simple property that may have this effect: the shortness of the individual's life span. I have previously[7] posed the question of whether this is the reason for the phenomena of aging and natural death.

That organisms can die a natural death is indeed not a matter of course. Biologically speaking, this at first does not seem required. The growth of a species can always be checked by enemies, hunger, and illness. In the majority of cases, these factors may well be the decisive ones: were all rabbits to die only of old age, the world would soon be overrun by rabbits. Nor can one see a physical necessity—in the sense, say, of a 'wearing out' of the organism. For in any case the materials in an organism are significantly exchanged during its lifetime, and the controls need not become obsolete. If nature did impose a necessary limit on the regenerative capacities of organisms, why should that limit lie at a single day for the May fly, at 8 years for man, and at no discernible time for the constantly dividing protozoans? The processes of aging exhibit a norm, they are strictly law-like; think of the onset of age-connected farsightedness in man, or of menopause. This lawfulness admits of no other interpretation than the existence of a control mechanism that enforces aging in an individual, and that in the end requires death. Since all higher animals are, without exception, subject to these processes, one should try to understand them as part of the health of the particular species, even though they affect the individual like an illness.

The dying individuals make room for new individuals of the same species. In a given span of time, a short-lived species therefore tries out more mutants than a long-lived species (assuming all other factors to be the same). The short-lived species thus has a chance of developing more rapidly than the long-lived one. In

[7] von Weizsäcker (1949), lecture IX.

this way, long-lived species are gradually being eliminated. A selection pressure exists in favour of short-livedness; i.e., of the shortest span of life compatible with the rate of growth of the species and the time it requires for breeding. As it says in the Tobler-Goethe *Hymnus über die Natur* [hymn on nature]: "Death is [nature's] artful device for having an abundance of life."

In somewhat loose connection with the preceding, let me remark on the estimation of time scales for development. Various authors have tried to estimate the possible time needed for the development of certain species and organs (e.g., the vertebrate eye) by means of chance mutations followed by selection. In attempting to be precise regarding the individual necessary steps, they frequently arrived at time scales that greatly exceeded the 5 billion years available in earth history. On this basis some of them concluded that the Darwinian explanation of evolution is untenable. In principle, we can counter thusly: one can try to compare all feasible paths of evolution capable of leading from a specified state A to an equally well specified state B (e.g., from organisms without eyes to vertebrates with eyes). If one had a complete list of these feasible paths, one could write next to each possible path the span of time it requires according to a statistical estimation. These spans of time would vary strongly from path to path (i.e., by many powers of ten), since products of small probabilities frequently enter into the estimation. The estimated time span in these calculations is the expectation value of the duration of the total process along the path in question. The path actually taken in the history of the organisms will be the one of shortest duration; once organisms with eyes have developed along this path, they will be superior to their eyeless competitors and will eliminate them. If we have been unable to think of all possible paths, and especially of the actual path taken by the historical development, then all the time scales calculated for the paths one has thought of will be too long (too long by many powers of ten, probably). This reflection shows, first of all, the difficulty of the attempt to refute Darwinism by such estimations. Secondly, it shows the great significance of all controls advancing the rate of progress; insofar as such controls can develop at all, one may expect their full victory in the struggle for survival.

However, rapid development is by no means always an advantage, as shown by the blind alleys of overspecialization, or by the not uncommon progression from 'orthogenesis' (e.g., the development, within a species, of successively larger tusks or antlers) to 'giantism' (unfunctionally large tusks or antlers). We may surmise that progress cannot in principle be fully expressed in norms. The theory of ideas belongs not to open, but to cyclical time.

6.4 Good and Evil

We shall use the words 'good' and 'evil' in the narrower sense of norms for the behaviour of individuals in a community. Human communities demand certain modes of behaviour of their members. What is the basis of these norms?

Following the example of behavioural scientists, we shall speak of 'moral analogues' in animal behaviour; these we will consider first. The preceding considerations apply directly. The only difference is that the control system is no longer the single organism but the entire animal group. If we again limit ourselves at first to the preservation of the community, its health and the health of its members can be defined in analogy with the health of the organs in an organism. An individual is 'socially healthy' if its norm of behaviour contributes to the preservation of the community—more precisely: if its behavioural norm is a part-norm of the healthy behavioural norm of the entire group. The Platonic state, incidentally, is organized precisely according to this principle. The pre-eminent social virtue, justice, consists of everyone 'performing his own task'. To Plato performing one's own task of course does not primarily mean—as in his slightly ironical examples—that everyone becomes specialized, that the shoemaker only makes shoes, but rather that each member acts in accordance with his idea; and this idea *is* the healthy norm. That is why, for example, the Platonic philosopher does not specialize, but—in a society healthy enough to allow this—at the same time *is* a statesman. For the statesman performs his own task only if he thinks the truth. To be sure, this requirement transcends the level that can be illustrated using the example of animal societies.

The definition of the good in terms of the well-being of society characterizes the morality of group self-preservation. How many subtle controls can serve this end is shown, for example, by Lorenz's analysis of aggression. Of course we then also get social illness; i.e., automated faulty social behaviour. The unbridled egotism of the individual is a social illness in many respects. In the same way, the unbridled egotism of a group can signify a social illness with respect to the preservation of the species.

In the case of man, we no longer speak of the moral analogues of behaviour, but of morality itself in the dual sense of the norms of traditional customs and of the norms of conscious decision. Let us here deal only with the latter; the highest level of the principle of such a morality of preservation is, it would seem, Kant's categorical imperative, which commands that the maxim of my action should be universally applicable. In the language chosen here, Kant demands: Will possible norms! 'Possible' means what previously was meant by 'healthy'. Here, too, group egotism can be a norm of illness just like private egotism. Of course reference to the biological species man is lacking in Kant, but then even in Darwinism this conception is useful only within limits. Kant's totality is the totality of reasonable beings; i.e., the very beings capable of conceptualizing a norm. Within the framework of the cybernetic model, this points back, on the one hand, to the principle of progress that transcends the limits of the species, and forward, on the other hand, to the problem of true thinking; i.e., of truth and falseness.

Already in the case of animals, the principle of progress relativizes the relation between health and illness also in the social domain. Ritualized aggression, which for Lorenz is the basis of the individual ties among individuals, and indeed the conception of ritualization altogether, proves that what is ill can, in a certain sense, transform itself into what is healthy in some new order—i.e., it proves that 'so-called evil'[8] can be transformed into the good. In this instance also, no automatically acting norm of progress can be named. The historical success of an innovation cannot be guaranteed. In the case of man it is thus doubly important that we try to grasp the unique in its uniqueness. Here we encounter the structural reason for the superiority of the reality of love over the principle of justice. But one cannot speak meaningfully about all this until one asks what truth is.

6.5 Truth

Truth is traditionally defined by a formula such as *Veritas est adaequatio rei et intellectus*. This formula provokes questions that are best articulated in the attempt to translate this statement into English. If we say something like "Truth is the correspondence between object and understanding", then truth appears as a relation called a 'correspondence' between two things; i.e., the object and the understanding. I say, for example, "there is a cat," or "2 is a prime number." How can the object—in this instance, the cat or the number 2—correspond to my understanding? My understanding, after all, is neither a cat nor a number.

One can restrict truth to the so-called truth of judgment, in which a proposition and a state of affairs are supposed to correspond. As Heinrich Scholz used to say: "The proposition 'Mars is inhabited' is true if and only if Mars is inhabited." This formulation clarifies the problem. If one takes Scholz's formula as it is written here, then two of its statements do correspond to each other—namely, the statement in quotation marks, "Mars is inhabited," and the subsequent conditional clause, "if Mars is inhabited". Let me not discuss the precise meaning of 'to correspond'; in any case, it will be readily admitted that two statements can correspond to each other. But where does that leave the definition of truth? Clearly "Mars is inhabited" represents the statement whose truth is to be examined; "if Mars is inhabited" reveals the condition for correspondence with the possible state of affairs, "Mars is inhabited" is the possible state of affairs itself. Thus the first statement, "Mars is inhabited," is true precisely when the second statement aptly renders the real state of affairs. The latter, however, is obviously what one means in saying that the second statement is true. I.e., a statement is true if it corresponds to a true statement; a true statement, however, is one that aptly renders the real state of affairs. What we have thus won is a new way of circumscribing the Latin definition of truth: *intellectus* is replaced by 'statement'; *res* by 'real state of affairs'; *adaequatio* by 'apt rendition'.

Of course, Scholz, who was here following Tarski, clearly understood the situation; his proposition was in fact intended to make the situation conspicuous. Tarski's solution of the problem, designed for formalized languages, lies in the

[8] Allusion is here being made to a work by K. Lorenz whose German title, translated literally, is 'The So-Called Evil'. The title of the English translation is 'On Aggression'—*translator*.

concept of meta-language. For a formalized 'object language', truth can be defined by means of a meta-language capable of saying everything that can be expressed in the object language. For meta-linguistic propositions, truth is defined either not at all or else by means of a higher meta-meta-language. In this procedure, a last recourse to 'natural language' remains inevitable. What we are concerned with, however, is the meaning of truth in 'natural language'. It is obvious that we cannot express this meaning with the precision that mathematical logic demands of itself. We are satisfied with seeking a direction in which we can continue asking meaningful questions.

Have we gained anything with the new formulation, "Truth is the apt rendition of a state of affairs by means of a statement"? Hardly; we have instead escaped from the demanding rigor of an *adaequatio* or 'correspondence' into the very vague formula of an 'apt rendition'. How can a linguistic construction such as a statement 'render' a state of affairs? After all, this is precisely what was supposed to be clarified. It would be simple if we could 'exhibit' a state of affairs also by some means other than a proposition, and then compare it to the proposition. Exactly this, however, seems never to succeed. If I point my finger at a cat, it is unclear whether I intend the state of affairs "there is a cat," or, for example, "Mimi is back home," "the cat is wearing a collar," etc. Furthermore, the state of affairs "2 is a prime number" can hardly be presented other than in a linguistic context. Perhaps states of affairs on principle cannot be presented otherwise than as propositions,[9] and the philosophical suspicion is aroused that states of affairs exist only for beings capable of language. If the correspondence theory of truth should fail now on account of this situation, how will we at all be able to distinguish truth from falseness?

The correspondence theory had its origin in the very Greek doctrine of eidos to which we referred in our explanation of health. According to Aristotle, speech (logos) is true if it connects what is connected in things (pragmata) and separates what is separate in them.[10] Speech declares the experiences of the soul (pathemata tes psyches). The soul recognizes something when it contains the same eidos as the thing. This is a form of the ancient doctrine that like is known only by like. In this doctrine we have exactly the required two 'objects'—namely, the eidos in the soul and the eidos of the thing—and these two may indeed correspond or not correspond, since both are of the same kind—namely, eide. The problem now shifts to the question of how a thing "can have an eidos," and how the same eidos "can be in the soul." It shifts, in other words, to the understanding of the meaning of the eidos doctrine. This is already an important step; it points up the sense of the thesis that logic depends on ontology for its final justification. For us, the question is now whether the cybernetic model of the eidos doctrine can help us along.

---

[9] See Patzig (1964).

[10] The categorical judgement, which assigns or denies a predicate to a subject, serves as the model for 'speech'.

Let us first return to animal behaviour. The young songbird in his nest 'recognizes' the mother who brings him food in her beak. He will also 'fall for' a dummy and for it, too, will open his beak. A chicken sitting under a bush within reach of a sprinkler, dripping wet on a sunny July day, 'misinterprets' the water of the sprinkler as rain. An animal can behave 'rightly' or 'falsely'. As an experiment, let us substitute this rightness for truth in the definition. We thus say: "Rightness is the adaptation of behaviour to the circumstances." Right behaviour, in the previously defined sense, is healthy behaviour.

The new formula advances us a step by replacing correspondence, which can be conceived of only between things that are alike, with adaptation, which is possible precisely between things that are unlike each other. Behaviour is certainly not an 'image' of the circumstances. It fits the circumstances not as a photograph fits its object, but as a key fits its lock. The monad that is an individual animal mirrors the universe of circumstances by means of the pre-established harmony of its inborn (or perhaps learned) behaviour; to be sure, this monad has windows. In all that follows we will try to hold on to adaptation as the interpretation of *adaequatio*. At the same time, we permit ourselves the stylized use of the term 'truth' for the Rightness of animal behaviour.

If truth thus aligns itself with health, then falseness must align itself with illness. In the domain of falseness one must of course differentiate between the quickly corrected error, the stubborn error, the lie, etc.; again, this will not be done here but will merely be presupposed as possible from case to case. After all, we have from the start elucidated health and illness in terms of correctly or falsely established control values. In order to be able to define health and illness, we had to first appeal to an understanding of truth and falseness. We now reverse the direction of the question, so as to get a better view of the presupposed structures.

Nietzsche calls truth the kind of error without which a particular kind of living being could not exist.[11] The shocking apparent logical circle in this sentence ("What is error, if truth itself is a kind of error?") acquires a positive meaning in our context. We first turn the sentence about and say: "Error is that truth without which a living being could no longer live." In the mode of expression we have been using: error is an illness; i.e., a faulty adaptation that is not immediately deadly. A faulty adaptation, too, is an adaptation; thus error, too, is a truth. In the theory of ideas: falseness, too, provided it can be articulated at all, contains something like an idea. But the erring adaptation is incomplete, it is 'inadequate' (*adaequatio*!). The erring organism imperfectly mirrors the environment that is its universe.

Actually, a fully adequate adaptation of an organism to the world does not exist. Every behavioural pattern can be deceived; i.e., if measured by rigorous standards, every behavioural truth is still an error. Thus Nietzsche's saying is exact: truth is an adequately well adapted error.

Even in this form, to be sure, the saying presupposes a certain familiarity with 'world', 'organism', 'adaptation', and it presupposes the sciences of biology,

---

[11] F. Nietzsche: *The Will to Power*, No. 493.

cybernetics, and physics. It presupposes these sciences—i.e., it uses their modes of speech as the correct scheme of behaviour; it presupposes their truth. In what way science can be true, the saying does not yet explain, however. Here the question arises what truth is for man, and thereby what truth is in its essence. With regard to this I will merely offer a hint, since it is actually a question of philosophy proper.

Man is restricted neither to modes of behaviour that are inborn nor to ones that are learned without understanding. He can act 'with insight', or 'with understanding'. In speaking of truth, we normally mean this 'insight' and 'understanding'. Frequently we even isolate the achievement of this capacity as 'theory', and only to it we ascribe truth in a narrower sense. Our approach, intended to scrutinize the background of this isolation, must now try to show which cybernetic model might represent the transition to this 'theoretical behaviour'.

The capacity of imagination appears to be an essential feature of insightful behaviour. I am not forced to perform one particular action; rather, I have several modes of action to choose from. I imagine the state of affairs and my action in it, I go through the gamut of possible events, and then I choose. Thus something like an "image in the soul of the state of affairs" appears here for the first time. Indeed, it was not to be expected that we would be able to bypass the traditional theory of truth. But we can now designate more precisely the meaning and the limit of the notion of an image. The image is a simulation, accomplished by consciousness, of what may happen to us in the world of events. It is true insofar as it is adapted to "actually possible courses of events" the way a key fits its lock. In the process of acquiring knowledge man of course reaches beyond any adaptation already achieved, beyond any 'lock' already existing. No statement of human knowledge has an unchangeable "ecological niche" in which it is definitely true. At the most it could be claimed that the true is the whole. But the whole is precisely what is not at our disposal.

A theory of truth that wishes to be true must ask whether its own truth falls under the truth it describes. Our theory is a model of truth. In the beginning, we defined a model as a mode of objectifying. Objectifying is an activity—namely, an activity that posits decidable alternatives in the flow of events. In accordance with our theory, our theory's truth is the adaptation of objectifying activity to the possibility of human life in the world. To this extent, our theory itself describes the condition for its truth. How this is possible, it does not describe. New questions would have to be raised at this point.

References

Gaiser, K., 1963: *Platons ungeschriebene Lehre* (Stuttgart: Ernst Klett).

Patzig, C. G., 1964: in: *Argumentationen, Festschrift für Josef König* (Göttingen: Vandenhoeck & Ruprecht).

von Weizsäcker, C. F., 1949: *The History of Nature* (Chicago: Chicago University Press).

Chapter 7
Parmenides and the Graylag Goose

Written in 1970 and unpublished until now.[1] The confrontation attempted in "Models…" (Chap. 6) between the species concept in contemporary biology and Platonic philosophy, which served to cast light on biology, here serves instead to elucidate Plato. We wish to show that Plato, in his critical analysis of the 'Ideas' in the Parmenides dialogue, speaks of the very problems that are also our concern.

Modern readers of Plato who try to understand what he meant by an Idea are sometimes led to a perfunctory insight: "Oh, I see, what he means is a concept." The complaint usually follows then that Plato turned concepts into pseudo-things. It seems vital to me that the contemporary scientist should realize his own incomprehension of what he means by a concept. The concept of a biological species serves as a fairly good illustration ("Models…" (Chap. 6) and Sect. 7.1 of this essay), but requires an excursion into logic (Sect. 7.2). We learn to understand that Plato's theory of Ideas is an attempt at showing what we ought to mean by a concept. If there is to be a meaningful cybernetics of truth, it should be possible to describe it also on the level of cybernetics itself. Section 7.4 ends in establishing this connection, but stops short of the topics that would lead us back into my "Matter, Energy, Information".[2]

---

[1] This text was first published in: *The Unity of Nature* (New York: Farrar Straus Giroux, 1980), pp. 357–378, is a translation by Francis J. Zucker of: *Die Einheitder Natur* (Munich: Hanser, 1971): IV.5.

[2] The text "Matter, Energy, Information" is reprinted as Chap. 11 in: Michael Drieschner (ed.): *Carl Friedrich von Weizsäcker: Major Texts in Physics* (Cham et al.: Springer-Verlag, 2014) and was originally published in: *The Unity of Nature* (New York: Farrar Straus Giroux, 1980), pp. 274–294; it was translated by Francis J. Zucker from: *Die Einheit der Natur* (Munich: Hanser, 1971): III.5.

M. Drieschner (ed.), *Carl Friedrich von Weizsäcker: Major Texts in Philosophy*, SpringerBriefs on Pioneers in Science and Practice 23,
DOI: 10.1007/978-3-319-03671-7_7,

7.1 What This Essay Is About

In the last Book of *The Republic*, Plato introduces what we may call the “Ideas of things”. His examples are the couch (*kline*, on which one reclines at table) and the bridle (reins and bit). Both implements (*skeue*) are discussed on three levels, each being a representation (*mimesis*) of the one directly above it. There is the Idea of the implement, there are the implements we can see and touch, and there are pictorial representations of such implements. Taking the couch as an example, Plato shows that each of these three is produced by a master craftsman: the Idea of the couch by God, the actual couches by a cabinetmaker, the pictures by a painter. Taking the bridle as another example, he shows that three skills pertain to it: the skill of the rider who uses the bridle in riding, the skill of the harness maker who makes bridles, and the skill of the painter who paints them.

How are we to understand the parallelism between God and the rider which is implied here? I owe the following interpretation to a remark made to me by Georg Picht: the rider, in riding, knows that structure of the world which enables horse and man to be in the relation we call riding, and he knows in what manner this relation is made possible by the bridle; in other words, he knows the function of the bridle. This function is in turn made possible by the bodily structure and psychological endowment of the living organisms, horse and man (both well adapted to their environment). According to the myth in the *Timaeus*, to which Plato here refers us, this structural whole was made by the Creator-God. It is He, therefore, who created the possibility of the function of the bridle, and this possibility Plato terms the Idea of the bridle.

Konrad Lorenz offers an analogous example from modern science. The graylag goose described by the zoologist and behavioural scientist is never encountered in reality exactly as it is described; i.e., as an ‘ideal type’ with regard to bodily structure and pattern of behaviour. But science refers to, and is indeed possible only with a view to, this ideal type. According to Lorenz, the reason for this lies in the Darwinistically interpreted environmental conditions of the animal. With its bodily structure and behavioural pattern, the graylag fits into a particular ‘ecological niche’. The ideal type describes a goose that is optimally adapted to this niche; around it, the characteristics of real geese are spread in a distribution dependent on various external parameters. We contend, in accordance with Picht’s interpretation, that the term ‘ideal type’ refers, in a functional sense, precisely to the Platonic Idea of a thing—i.e., in this instance, of the graylag.

Lorenz’s account can be supplemented in an important respect. (I am greatly indebted to a correspondence with Otto von Helversen on this point.) The ecological niche does not completely define the type ‘graylag’.[3] Rather dissimilar

[3] Biologists have talked to me on several occasions of the oversimplified presentation of the species problem here and in “Models…”. I have evidently failed to point out the context in which this presentation is meant to be understood. Although in constant use, and thus clearly pragmatically useful, the species concept cannot, on the other hand, be justified theoretically. The

organisms can fit into fairly similar niches; compare, for example, the highly specialized Darwin finches on the Galapagos Islands and marsupials in Australia with the corresponding animals on the large continents. A geneticist would say: the chromosomes define the phenotype of the individual (up to individual fluctuations), and are in that sense a physical realization of the genotype, which can be considered the 'Platonic Idea of the phenotype'. In a further approximation, selection within an ecological niche picks out one of the many possible genotypes as 'the best', the Platonic Idea of genetically closely related genotypes in that niche. Though still too short, these remarks point out the complications into which one is necessarily led in the detailed pursuit of these thoughts. The complications are indeed unavoidable if the basic hunch is correct that the Platonic Ideas of things are the structures in the world that make these things possible: for these structures are in fact complicated. But how valid is this hunch; i.e., what does it contribute to the interpretation of Plato's philosophy, and what to the philosophy of modern science?

The Ideas of things do not at first appear to be a suitable point of departure for the study of Plato's philosophy; an analysis of the sort we have just carried out can at best remove the psychological obstacle initially in the way of such a study (as it is usually conceived); this obstacle lies in the uninformed opinion that Plato's theory of Ideas is outdated and no longer relevant. Plato himself, in the disposition of almost all of his writings, as well as in the chronological order of his entire oeuvre, takes ethical and political problems as his point of departure and quickly progresses to the central problems of Being, Knowledge, and the Good. The Ideas are introduced as a step in this progression—and it is immediately and in a quite natural manner the highest Ideas that are thus introduced. Mathematical examples serve as models for discussion, but the main emphasis is on Ideas such as the Same, the Just, the Beautiful. A further step consists of the reflection on what has thereby been accomplished. In this critical reflection the preceding step appears as the 'Idea hypothesis', while it is only the reflection itself that introduces the 'theory of Ideas'; i.e., the doctrine of what constitutes Ideas. (In this methodological distinction I follow the terminology of Klaus Meyer-Abich.) The question of what can and cannot be Ideas must form part of the theory of Ideas, and it is here that the notion of the Ideas of things necessarily has its place. It seems that only within the framework of a fully developed theory of Ideas can one state what this notion means.

(Footnote 3 continued)
definition of a species as a class of interbreeding organisms is sometimes not adhered to (as in the case of plants, where finer subdivisions are still counted as species); the definition is also insufficient in principle, since among three geographical varieties A, B, and C, A may be able to interbreed with B, and B with C, but not A with C. The concept is therefore non-transitive and cannot serve as a basis for classification. Why, nevertheless, is the species concept so useful? It is here that the pointer to the theory of selection implied in the concept of the ecological niche serves as at least a rough explanation. I presuppose the feasibility of such an explanation here, as I did in "Models…", and do not wish to claim more. The Platonic 'Ideas of things' must, in accordance with Plato's own teachings, show a vagueness similar to the species concept's.

By stylizing the circuit through Plato's philosophy as a projection of the Parable of the Cave onto natural science, it should nevertheless be possible to choose the problem of the Ideas of things as one's point of departure on this circuit. Putting it in terms of texts, this implies an approach to the relation between the three Dialogues *The Republic*, *Parmenides, and Timaeus.* The prisoners in the cave take the shadows on the wall for reality. (I am here following Picht's explication of the parable, but in a rather free interpretation which is more restricted in scope, and for which I alone am responsible.) Projected onto science, the shadows on the wall are the ways in which the things of the world show themselves to our senses. The turnabout of the entire soul, with which philosophy commences, is the realization that even the physically real is already distinct from its appearance to the senses. The turnabout is thus the beginning of the search for truth—for what, in other words, the Real might truly be. The successive stages in which this search reveals the things in the cave, the mirror images beyond the cave, the things beyond the cave, and the light of the sun itself stand for the stages in the successive reflections on what revealed itself as the Real in each of the preceding stages. In the course of these reflections the restriction with which we started must dissolve of itself, since the possibility of restricting oneself to natural science is itself merely a mirage. In the Parable of the Cave this ascent is followed necessarily, as an ethical requisite, by the descent back into the cave. In our projection the descent corresponds to deductive natural science, which the *Timaeus* presents in half-disguised form. Only in the completed descent can the true nature of physical things reveal itself; and the term for the 'true nature of a thing' is its Idea. Thus the answer to the question with which we began the ascent must in the nature of things lie in the final stage of the descent. The ascent can be tersely sketched in the following sequence of statements: "What is this child playing with?" A ball. "What is a ball?" A sphere as presented in sense perception. "What is a sphere?" A mathematical Idea. "What is an Idea?" A true and good One. "What is the One?" Read the Parmenides Dialogue!

Let us follow the challenge of the preceding sentence. As Socrates tells us in the *Phaedo*, Anaxagoras had undertaken to explain everything in terms of Mind (*nous*). But he did not remain faithful to this undertaking. In his detailed explanations he reduced the appearances in the context of physical causality to principles which themselves are of the type of appearances. A group of philosophers from Clazomenae, the home of Anaxagoras, now arrive in Athens to inform themselves of the earlier colloquy between Parmenides and Socrates concerning the One—that is, concerning the highest principle of all possible explanation. They meet Plato's brothers Adeimantus and Glaucon, Socrates' interlocutors in *The Republic*; in this way the author gives a hint—just as in the background narrative of the *Timaeus*, only more subtly—that problems dealt with in *The Republic* are to be taken up once again. The brothers conduct the Clazomenians to their half-brother Antiphon, who once had memorized that colloquy but is now, as befits his social standing, interested only in horses. Immediately after handing a bridle over to the saddler for repairs, Antiphon turns to his guests. As Picht remarks, this is a hint that the Dialogue is to repair what the bridle represents. And indeed the colloquy that Antiphon then relates consists of two parts, which one can call the prelude and the fugue: the prelude criticizes the Idea hypothesis, i.e., it is an introduction to the theory of Ideas, and the fugue treats the basic concept of the theory of Ideas, the One.

The prelude merely asks questions. Answers are given neither in the prelude nor in the fugue. Had Plato considered these questions to be unanswerable, we would expect him to abandon the theory of Ideas, at the latest, in this text. Nothing in his work nor in the tradition supports this interpretation. (Even if it were true that the 'friends of Ideas' in *The Sophist* state an earlier view of Plato's, which I doubt, this Dialogue still contains only a reformulation of the theory of Ideas, not its renunciation.) We must therefore conclude that Plato requires the reader to find the answers himself, or perhaps in other Dialogues. Parmenides' assertion that one must do exercises before searching out the answer (the fugue being an emphatic example of this) strikes us as immediately convincing. If every Idea is true, existent, good, and one; if truth and being radiate from the Good as light and growth radiate from the sun; if the Good is to be equated with the One, then, in the final analysis, the question "What is an Idea?" leads back to the question "What is the One?" This question is of a special kind, since there cannot be any further recourse to an X in the form of a predication "The One is an X." It would be absurd to imagine that one could explain what an Idea is to someone who has not thought through this problem of the One; he would be unable to understand whatever one may say to him.

All one can do beforehand is to point the thoughts in a direction in which, as one proceeds, more and more will become evident. This is the philosophical function of all Platonic Dialogues. The interlocutor of Socrates (of the 'Eleatic', i.e., of Parmenides, respectively), who represents the reader, starts out with the kind of knowledge one usually has. This knowledge is undermined and, to the extent that the Dialogue is didactic, also explicated by invoking a higher systematic knowledge that, though perhaps unclearly, is available to the reader or can be awakened in him. In order to point out the necessity of further and still deeper questioning, this higher knowledge is frequently undermined in turn in the concluding paradoxes; nevertheless, it serves a positive function in the main body of the text. This entire procedure is made possible by the structure of the 'realm of Ideas', which invites a progressive ascent.

In what follows I attempt to write, though not in dialogue form, a Platonic Dialogue for modern scientists. The initial knowledge, which is to be criticized and explicated, is represented by a simple, seemingly unproblematic concept; for this purpose, and in order to link up with Lorenz, I choose the example of the graylag goose. The higher systematic knowledge invoked by the criticism and the explication is the general conceptualization of contemporary science. It will be shown that the concept under consideration has a place in this conceptualization that can be precisely characterized in terms of Plato's description of an Idea. This is accomplished by applying the questions in the prelude of the *Parmenides* to the Idea 'graylag goose', and by answering them in terms of the conceptualization of natural science. That Plato intended 'Ideas'—or even the special Ideas of things—in precisely this sense, is of course not being argued here. To this end, one would first have to subject the conceptualization of natural science itself to a critique, in order to then link it to the critique of all conceptualization in the fugue of the *Parmenides*.

7.2 The Idea of the Graylag Goose

I suggest the reader now turn to the text of Plato's *Parmenides* and read its first part, the prelude (126a–135c), allowing the following reflections to serve as a commentary.

The brief introductory narrative (to 127a) has already been discussed. The Dialogue next turns to Zeno's work containing the famous paradoxes of movement, on which we need not dwell here. Socrates, introduced as a gifted young man, points out that these paradoxes could be avoided by distinguishing the objects of perception from their Ideas; this, too, we will bypass. Parmenides praises Socrates for his good start in philosophy. The discovery of the theory of Ideas is imputed to Socrates (130b), while at the same time it is made clear that Parmenides and Zeno have long since been familiar with this theory, as well as with its difficulties. Our commentary begins with this passage.

First, a question concerning the historical situation. One is taught in the history of philosophy that Plato invented the theory of Ideas, which had occurred neither to the historical Socrates nor to the historical Parmenides. Plato's crediting both with this invention is supposedly only a literary device. If we accept this view, which in a literal sense is certainly correct, then the question arises why Plato chose this disguise.

We attempt to answer this question by means of a principle of interpretation (to be frequently applied in what follows) that can only be justified by its success. Let us call it the principle of the truth of what is asserted. One can put it this way: every assertion that Plato assigns to one of his characters has an interpretation in terms of which the assertion is true in Plato's own view. The misconceptions which at first appear, and which Plato himself, in the course of the Dialogue, frequently analyses or at least brings into the open, are meant to be recognized by the reader himself, so that all inadequate interpretations of the topic under consideration are gradually peeled away. That interpretation of an assertion that remains true to the last is also its interpretation in the sense of Platonic philosophy. The work of peeling away erroneous or incomplete interpretations is at the same time the work of ascending to the principles of philosophy. This work corresponds precisely to the structure of the theory of Ideas itself. Every assertion submitted to discussion in a Platonic Dialogue is a representation (*mimesis*) of truth; this truth is itself laid bare in the work of interpretation. This is an ancient exegetic principle of interpretation. I owe its sharp formulation to the *Laches* Commentary, unfortunately unpublished, by Georg Picht. Picht shows that the definitions of courage given in that Dialogue by Laches and Nikias literally refer to the definitive formulation of courage in the fourth book of *The Republic*, and derive their essential meaning from there. For the purposes of the present investigation, I merely propose using this principle as a hypothesis.

If this principle is applied to our historical question, then an interpretation of the theory of Ideas must exist according to which it would be correct to say that Socrates and Parmenides were already in possession of this doctrine. In a more formal sense this is almost self-evident. If the theory of Ideas is the true philosophy, and if Socrates and Parmenides, the two teachers most revered by Plato, were true philosophers, then they must have known the core of what the theory of Ideas deals with. But how should we articulate this core? This is precisely what we cannot accomplish, according to Plato, before having thought through the whole of the *Parmenides*. At the present stage we must make do with hints.

Aristotle reports that Socrates posed the question of what this or that (for example, justice) essentially is (*ti esti*), while Plato turned the answer to this question into something existing on its own; namely, to the Idea. In a banal version of the history of philosophy, this is made to mean that Socrates invented the concept, while Plato hypostasized the concept to an Idea. One should say, rather, that Plato thematized the question "What is a concept?" For it is by no means clear what terms such as 'concept', 'name', 'justice', and 'the essence of something' actually signify. If one is unwilling to think of Socrates as muddleheaded about his own procedure, one must credit him with having already asked himself what his procedure actually amounts to. But, in a certain sense, this means crediting him with a knowledge of the theory of Ideas—provided we believe, as Plato has him say in the *Meno*, that one can ask only what in a certain sense one already knows. Assuming that we were on the right track in section a, the answer to this Socratic question can be given only by invoking the One. If we believe that Parmenides knew what he was saying when he spoke of the One, then Parmenides had the key to the door whose knob was clasped by Socrates. For a discussion of what Parmenides actually taught, I refer the reader to an article by Picht.[4]

We now turn to our subject and impose on ourselves the restriction of treating it strictly in accord with the viewpoints of contemporary science.

Parmenides formulates the question as follows: "And now tell me: have you yourself drawn this distinction you speak of and separated apart on the one side Forms themselves and on the other the things that share in them? Do you believe that there is such a thing as Likeness itself apart from the likeness that we possess…?" (130b 1–4).[5]

---

[4] See Picht (1960).

[5] I quote Plato in the translation by F. M. Cornford: *Plato and Parmenides* (Atlantic Highlands, N. J.: Humanities Press, 1964), unless otherwise noted. Although, following Cornford, *eidos* is rendered as 'Form' in the quotations from Plato, I continue referring to Plato's theory of *eide* as his 'theory of Ideas' (in conformity with Weizsäcker's *Gestalt* and *Ideenlehre*, respectively). The remaining sentences in this paragraph (p. 450 of *Die Einheit der Natur*) are omitted here; Weizsäcker explains in them what guidelines he follows in translating the Greek text into German. To the extent that these guidelines apply also in English, Cornford's text already satisfies them (translator).

The rhetorical emphasis of this question lies in the thrice-repeated *choris* ('separated', 'apart'). The Forms are something apart from the objects that have this Form, indeed the Form is something apart from that which the object has in itself of this Form. This is precisely what Lorenz means by the ideal type of the graylag. The ideal type is neither one of the many graylags, nor is it the 'graylagness' that an empirical graylag has in itself; it is not even a statistical mean of empirical graylags. But what then is it? This is what we need to formulate. In principle, we already know the answer. The ideal type is the lawfulness that makes graylags possible. Let us therefore return to the text, be it to better understand Plato or to better understand the concept of lawfulness.

Parmenides first asks of what Socrates is willing to posit Forms. He lists examples that fall into three groups:

- the Just, the Beautiful, the Good,
- man, fire, water,
- hair, mud, dirt.

The list is incomplete—lacking, for instance, even the mathematical concepts that had already been mentioned. For the young Socrates, the list is a sequence of descending evidence. The august values in the first group are without doubt themselves Forms. Man, constituted as a physical being by the elements, is an example of the problematic idea of a natural science based on the Forms. Socrates feels helpless in the face of the third group, which is just as much to be rejected from the point of view of values as accepted from the point of view of natural science. Parmenides consoles him with the promise of that freedom from prejudice which the philosophical maturity still ahead of him will bring. The prejudices of contemporary natural science lie in the opposite direction. Science considers the Just, the Beautiful, and the Good to be subjective, while it has no difficulty in conjoining the Form of the feather to that of the graylag, the Form of hair to that of man, the Form of mud to that of the elements. (Dirt is sometimes defined as matter in an unsuitable location; thus dirt exists only in relation to a value system.) In Darwinism, values are replaced by the objective criterion of survival. Our present self-imposed restriction precludes a critique of this approach. We will therefore take up groups 2 and 3.

What, then, is a Form? Three suggestions are offered: a Form could be in things (*en hekasto*), it could be a thought (*noema*), or it could be a pattern (*paradeigma*). All three suggestions are refuted, and the discussion ends in a paradox. Three questions must be asked of each of the suggestions: (1) What, naively interpreted, does it mean? (2) Why is it mistaken if thus interpreted? (3) How, in accordance with the principle of the truth of what is asserted, can the suggestion be defended?

7.3 The Form in Things and as Thought

"You say you hold that there exist certain Forms, of which these other things come to partake and so to be called after their names: by coming to partake of Likeness or Largeness or Beauty or Justice, they become like or large or beautiful or just?"

(130 e4–131 a2). A comment on the translation: the term 'things' is not in the original text; rather, the vague plural neuter *ta alia* appears there, which could be rendered with a phrase such as 'all these others'; nothing is here being prejudged with respect to 'thingness' in a narrower sense. 'Partake' is *metalambano*, and the corresponding noun *metalepsis* I will translate as 'a partaking'. In these terms it is the preposition *meta,* which is so important to Plato; *meta* also appears in the terms *metecho* and *methexis*, which are used in parallel with the terms just mentioned and are usually translated as 'participate' and 'participation'. *Meta* should be translated by the preposition 'with'. That which participates in the Form 'has the Form with it'. Aristotle criticized this *meta* as a mere simile, as a metaphor lacking conceptual sharpness. He uses instead the preposition *en*—in English, 'in'—which he considers conceptually sharp, and on which Plato will soon comment in our text. 'To become like' is meant to render *homoia gignesthai. Gignesthai*, however, as Picht has frequently emphasized in conversations with me, is not simply 'to become' but also 'to appear', 'to manifest itself'.

The goose Martina shows herself to be a graylag insofar as she partakes of the Form 'graylag'. This expression will appear strained to the modern reader, as it probably also did to readers in Plato's time. Does it say anything at all? With the question "Do you hold, then, that the Form as a whole, by being *one*, is in each of the many things—or how?"[6] Parmenides entices the young Socrates onto the thin ice of a first interpretation. Let us for the moment set aside the special question of the wholeness and unity of the Form, which is here being thematized. The notion that the Form is 'in' the participating thing is introduced here almost as a matter of course. I will not follow up in detail the difficulties that Parmenides deduces from this conception (to 131 e7). The difficulties appear to stem from the tacit presupposition of a 'substance model' of Form—as though a thing were 'made big' because 'bigness' enters into it like a kind of matter (caloric, virus). Socrates offers the apt comparison of the Form to the day, which is one and the same simultaneously everywhere. Parmenides shows that the comparison has not been thought through by in turn comparing the day to sailcloth spread over sundry things, each covered by a different part of the sailcloth. If the being-in-the-things of the Form were meant in this sense, then there would have to be as many Forms of the graylag as there are graylags, and the zoological term 'graylag' would be meaningless. But in what sense can we speak here of the truth of the assertion under discussion? We will be able to discover this truth only after having tried to escape it.

A modern reader here encounters a somewhat different difficulty than did Plato's contemporaries. The Greek philosophers had to draw the terminological means for expressing highly abstract structures from the living flow of everyday speech, and to do so they had to reflect on the essence of the matter. The result of their labour has been preserved for us in the form of established scientific terminologies. We, conversely, must first make clear to ourselves how little the

---

[6] Here I follow Weizsäcker's, not Cornford's translation, which, it seems, misplaces the word 'things' (translator).

meaning of these terminologies should be taken for granted. Thus we would like to replace the somewhat 'laboured' sentence "The goose Martina shows herself to be a graylag insofar as she partakes of the Form 'graylag'," by the simpler "The goose Martina has the Form of a graylag," or, still more concisely, "Martina is a graylag." We are therefore dealing 'simply' with a predicative sentence, containing a subject and a predicate. According to this view, Plato's Forms express nothing but the grammatical fact that a predicate can be made the subject of a further predicative sentence: "The graylag is monogamous." Plato's difficulties are therefore due merely to his looking for ontological depth behind those simple forms; this is excusable, since he was still wrestling with the formulation of logic. The *Parmenides* is thus an investigation from the early history of logic. For us, however, its problems have been solved, since we know that propositions can take the subject-predicate form.

If problems are made to disappear in this manner, they show up sometime later in a more virulent form, like a bacteria strain grown resistant to penicillin. Linguistic relativism, a clever contemporary doctrine, discovered that the subject-predicate relation, that fundamental figure of Aristotelian logic, stems from a grammatical form developed in the Indo-Germanic languages. Other languages make use of entirely different structures. Therefore, our so natural conviction that we know what the sentence "Martina is a graylag" says results from a linguistic habit. Doesn't all talk in an Indo-Germanic language that tries to escape this appearance of obviousness turn in circles in the linguistic prison yard to which our tradition confines us?

I have dealt with this question in another essay.[7] Anyone who comprehends and can state what the linguistic relativist means is speaking about language in an empirical language, and in this manner about languages other than his own. In describing the prison, he has already left it. This possibility is related to the fact that if one speaks meaningfully about language one is already speaking, by means of language, about the things with which language deals. The things, with which language deals, however, are precisely what Plato calls the Forms. To understand language as a performance, and to understand the theory of Ideas, is therefore basically one and the same thing. This is exactly how Plato and Aristotle worked: they investigated language as an indication of Forms and did not hesitate to reshape language in accordance with the Forms they saw.

The function of linguistic relativism in the present context is merely to remind us that the meaning of the basic logical figures is by no means obvious. If we understand "graylag" as a predicate, then we must try to say what we understand by a predicate.

Predicates as such, it is usually said, do not exist. 'The graylag' does not exist, only graylags do. 'To be a graylag', just as 'to be grey', is merely a possible

---

[7] "On the Relativity of Language", in: *The Unity of Nature* (New York: Farrar Straus Giroux, 1980), pp. 64–71; it was translated by Francis J. Zucker from: *Die Einheit der Natur* (Munich: Hanser, 1971): I.4.

property of things. ‘Grey’, ‘graylag’, etc., occur as properties of certain things. Our linguistic expression again shifts the burden of the intended structure to a new preposition, the little word ‘of’. One might think that there is no essential difference between ‘of and ‘in’. The assertion that there are ‘things with properties’ is the very structure which the theory of Ideas tries to make explicit; and it does this in a way that stresses the possibility of meaningfully asserting something about properties without directly talking about the thing that has them. One need not know Martina in order to do zoology.

The ‘logical’ critique of the theory of Ideas claims, however, that this doctrine asserts nothing beyond the possibility of making a predicate the subject of a new sentence, and that this possibility is a matter of logic, not of ontology. We must examine this objection, and in order to do so we must ask its proponent what he means by logic.

One of the classical definitions designates logic as the science of thinking, or, more exactly, as the science of the universal figures of thinking. If this definition is to be relevant to our problem, then the critic must take the further step of saying that we need not know in what physical or ontological sense a property is ‘in’ or ‘of’ a thing. It suffices to know that we can think and speak about things and their properties in the forms of classical categorical judgments. Supposedly, the prob lem wrongly ontologized by the theory of Ideas is a problem of the structure of logical thinking, and all the seeming paradoxes of the Forms can be resolved within this framework. This is the way out attempted by Socrates in our text (132 b3-6): “But, Parmenides, said Socrates, may it not be that each of these Forms is a thought, which cannot properly exist anywhere but in a mind. In that way each of them can be one and the statements that have just been made would no longer be true of it.” And indeed, when I think ‘graylag’, I think but one thought, which nonetheless refers to many graylags. The unity of the Idea would thus be the unity of thought.

Parmenides refutes this objection with remarkable alacrity: “Then, is each Form one of these thoughts and yet a thought of nothing?—No, that is impossible.—So it is a thought of something?—Yes.—Of something that is, or of something that is not?—Of something that is. —In fact, of some *one* thing which that thought observes to cover all cases, as being a certain single character?—Yes.—Then will not this thing that is thought of as being one and always the same in all cases be a Form?—That again seems to follow.—Doesn’t it then necessarily follow, said Parmenides, that the things, as you say, partake of the Forms? Or do you hold that everything consists of thoughts and thinks, or else do you hold that the things, though they are thoughts, do not think?”[8] (132 b7–c11). (In several instances, my translation is already determined by my interpretation, for example in the splitting up of the last sentence.) For the moment, let us ignore the alternative in the last sentence which begins with ‘or’, and which is itself subdivided in

[8] The last two sentences in this quote follow Weizsäcker’s translation, not Cornford’s, for reasons which Weizsäcker’s next sentence makes apparent (translator).

that it offers yet another alternative. What is Parmenides' counter-objection? In our current language, it would be: a logic that is to be capable of dealing with what is real must in its own structure mirror the ontological structure. Precisely if it is possible to think with the one thought 'graylag' something relevant to all graylags, then something relevant to all graylags must exist which this thought is thinking. It is this real and relevant something that is the object of the theory of Ideas.

This idea is compelling; but it runs counter to an intellectual habit that is especially widespread in our time. The idea asserts that it is altogether senseless to try to justify logic without at the same time doing ontology. And if logic is understood as the science of thinking, and thinking as the thinking of something, then there is no way out. One can look for ways out by interpreting either logic or thinking in some different manner.

In the last sentence, Plato suggests a way out that cancels the separation between thinking and what is thought. Logicians are bound to find this way out too objectionable to take it seriously, and since Plato himself does not pursue it, we may feel at liberty to drop it. But this is how Plato often sneaks in decisive hints. The horizon opened by this hint transcends the bounds of contemporary natural science, however, within which we now wish to remain.

How about a different interpretation of logic, then? I would like to discuss two variants: extensionality and the operative interpretation of logic.

A strictly extensional logic interprets properties as classes. The property 'graylag' is then simply the class of all graylags. This way of talking avoids the difficulty of the 'in'. The class of all graylags is not 'in' every graylag; inversely, one could say that each graylag is 'in' the class of all graylags. Every proposition on 'the graylag' is in truth a proposition on all graylags. Thus all difficulties of the theory of Ideas seem to disappear. The only 'ontology' one still requires is the thesis of the applicability of class logic to reality—i.e., the thesis of the (at least approximate) classifiability of reality into distinguishable individuals. 'The graylag' is then indeed a thought to which exactly one object corresponds; namely, the totality of all graylags. In anticipating the third interpretation of Form as pattern or example, we can ascribe, in accordance with the logical rule *dictu de omni*, propositions valid for all graylags also to an individual graylag; this graylag, as one sometimes says in mathematics, then serves as a representative of its class.

But this reduction of the theory of Ideas to logic merely transposes the problem to two other places: to the structure of logic and to the problem of the application of logic in temporal reality. In interpreting propositions on 'the graylag' as propositions on all graylags, I have been oversimplifying. In logic one must distinguish propositions on a class from propositions on the individuals in it; one also distinguishes between a class containing only one individual, and that individual itself. But this is Platonism; it is the separation of the Idea from the things which partake of it. If logic is unable to dispense with this distinction, it confirms rather than reductively explains the theory of Ideas. For the present, I am unable to delve more deeply into this matter; here lies a task for future work.

The notion that logic is not based on the psychology of thinking has, since Frege, gained universal acceptance. Logic does not state the principles according to which we do in fact think, but the principles according to which we ought to think if we are to think truthfully. Logic deals with structures of truth. While the extensional conception of properties appeared to offer an explanation of the theory of Ideas by means of logic, this anti-psychologistic turn of logic seems almost like an explanation of logic by means of the theory of Ideas. 'Structure' is a modern translation of *eidos* or *Idea*, and truth is the essence of the Idea; the structure of truth is therefore an aspect of the Idea of the Idea, i.e., of the One. Since all this sounds enigmatic to modern ears, one tries to disencumber oneself of the enigma.

Frege's logic is, to be sure, fundamentally an abstract logic. This logic has failed, and the operative interpretation of logic which, as a follow-up on Brouwer's intuitionism, is primarily represented by Lorenzen, appears a way out. Logic, it says, is a system of rational rules underlying the success of certain performances, be it as the system of rules admissible in every calculus or as the theory of successful strategies in a discussion game. The universal structures whose rationality is here being postulated are no longer structures of all objects of thinking; this is how the claim could arise that logic has been freed of all ontology. These structures are, rather, structures of possible performances, and thus, in the last analysis, structures of time. Here we have a further ground for the conjecture that the modern theory of Ideas will refer us back to time—as the functional interpretation of the bridle has already suggested to us. Indeed, I would not hesitate to designate the science of the structure of time as the ontology of science. Lorenzen stops short of this question, since he desires to exhibit the evidences underlying logic but is unwilling to investigate the ground of their possibility; on this point he seems like a Platonist who, while presenting the hypothesis of Ideas as self-evident, dispenses with a theory of Ideas.

The application of logic to temporal reality—for example, to the graylag—also shows how questionable the extensional conception of properties is. Science is empirical insofar as it predicts future experience on the basis of past experience. Science does this by stipulating universal laws which embrace the future events. A proposition on "the graylag" must also be valid for future graylags. But these are still unknown to us. A proposition on the class of all graylags therefore pertains to a class whose individuals on principle cannot be counted beforehand, even though nobody doubts that their total number is finite.

Only at this point—after a detour via logic that, with our contemporary consciousness, we probably cannot avoid—do we arrive at the meaning of the Platonic concept of the *eidos*—the Form—in natural science. How does one decide in the first place whether a particular animal is a graylag? It is impossible first to pick out from all animals a particular number of them as graylags, and then decide for Martina whether she is among those we have picked. Rather, it must be the other way round, we must specify characteristics that allow us to decide, when we encounter any animal whatever, whether it is a graylag. In science, the concept of property precedes that of a class—the more so if one considers that a sharp demarcation between classes is impossible. Is this mutant, this geographical variant still a graylag? It is well known that the characteristic of being able to produce viable offspring, which is meant to define membership in a species, is not transitive.

For the sake of comparison, let us glance at the Aristotelian conception that the Form is in the things—in the graylag, in our example—its eternal duration guaranteed by the circumstance that graylags always produce graylags. Within the approximation in which one can assert the constancy of a species, this way of talking is defensible. But in view of the theory of evolution, we need a concept of Form that makes the Aristotelian 'in' appear as a mere metaphor, while the Platonic 'participate' correctly represents the distance between the 'ideal type' and the empirical individual.

Nevertheless, this discussion has not yet solved but merely reformulated the philosophical problem of the meaning of Form and participation. We see the necessity of some sort of hypothesis postulating Ideas, but without having constructed a theory of Ideas. This is brought home by the objection that the Form 'graylag' is obviously not something in the things, but merely a thought useful in imposing order on things; the thought would be superfluous if we had an exact knowledge of things—for example, if we could trace in detail the interplay of the atoms in the body and environment of graylags. Two counter-objections may be cited.

First, the concept of an atom is, most likely, of the same nature as that of the graylag. The atom, too, is a Form approximately defined by the lawfulness of events. But to make this clear, one would have to think physics through with our question in mind (cf. "Parmenides and Quantum Theory").[9] One can say, in general, that the theory of Ideas is meant to explain the possibility of forming any concept whatever. That is why shifting from one conceptual level—for example, that of biological species—to another —for example, that of atoms or elementary particles—does not lead us to a point beyond the theory of Ideas, from which we could raise objections against it.

Secondly, it seems that the recourse to the movements of individual atoms tends to make us lose our awareness of biological events, rather than clarifying them. This is connected with the problems of the definition of the concept of information. Let us pick an example from within physics. Temperature is a statistical concept. The individual description of all atomic movements would cause this concept, and with it the phenomenon of irreversibility, to disappear altogether. Our experience, however, cannot be described without a conceptual formulation of this phenomenon. The situation is analogous when it comes to the concepts of genetics, which presuppose irreversible events. There are laws which manifest themselves to us in 'Forms' like heat conduction, the development of species, and heredity, and which one can recognize though one has not yet understood their reduction to atomic processes. Here, too, a specific formulation of the theory of Ideas appears to be needed.

The present essay is not meant to treat material problems of science but only the connection between science and Platonic philosophy. The question is: How do we understand in current terminology what that entity which we designate by the

[9] Cf. Chap. 8 in this volume, reprinted from von Weizsäcker (1980).

Platonic term 'Form' (or 'Idea') might be? Unless we can develop a new and clearer concept of "being in," this entity is only metaphorically 'in' the things. It is a thought, insofar as everything that can be known is thought by us in the knowing of it. But *what* is here being thought remains an open question.

7.4 Form as Example

"But, Parmenides, the best I can make of the matter is this: that these Forms occur in nature like examples, whereas the other things are like them, are their likenesses; and that the other things participate in the Forms in no other way than by being made in their likeness"[10] (132 d 1–4). Two words in this text pose a problem for the translator: *paradeigmata* and *physis*, which I render as 'example' and 'nature', respectively. Etymologically, *paradeigma* is something shown next to something else; in common usage, it is normally an example, often in the sense of a model. *Physis*, according to its root, is best rendered by 'growth'; in the Latin philosophical tradition it is translated as 'nature'. On recognizing the classical presentation of the theory of Ideas in this passage, one will paraphrase *paradeigma* as 'archetype' and translate it as 'prototype'. This is how the word *paradeigma* is used, for instance, in the introduction of the great speech of Timaeus (*Timaeus* 28 a 7). *Physis* one will translate as 'true nature', 'the nature of things' (thus Cornford in *Plato and Parmenides*, p. 93), or as 'the true Being'.

Parmenides' criticism, which follows immediately, evidently refers to this interpretation. Just as a likeness is like the Form, so must the Form be like the likeness; the relation of being alike is symmetrical. But two things that are like each other must participate in the same Form, and this Form is then the Form properly speaking. This argument can be iterated; i.e., the classical hypothesis of Ideas in the sense of a prototype-copy relation founders on the infinite regress of the 'Third Man' (what man as perceived and the Form of man are alike in, is the Form of man properly speaking, etc.). Cornford considers this argument fallacious, since according to Socrates' own assertion (129 a) two people are alike not by virtue of their participation in the Form man, but in the Form alikeness. This objection seems to me merely formal, because the hypothesis of Ideas developed out of the question of what two things of the same kind are alike in. Cornford is correct in remarking that Plato continued (e.g., in the *Timaeus*) to use the prototype-copy relation for the formulation of the Idea hypothesis, and that he therefore evidently did not consider it as refuted; but the question is how Plato, while continuing to use this relation, expected it to be interpreted. Parmenides' relevant conclusion is contained in the sentences: "If so, nothing can be like the Form, nor can the Form be like anything" (132 e 6–7), and: "It follows that the other things do not partake of Forms by being like them; we must look for some other means by which they partake" (133 a 5–6).

[10] This passage follows Weizsäcker's translation, not Cornford's, for reasons that the text is about to explain (translator).

The concluding 'greatest paradox' of Parmenides (133 b 4–135 b 2), which I will now merely sketch, strikes one at first as a grotesque overstatement. How can there be any relation between the two worlds of prototypes and copies, even a relation between knowing and being known? The human master (i.e., master-as-copy) is master of the human slave; the Master as such, i.e. the prototype of the master, is Master of the Slave as such, i.e., of the prototype of the slave. All relations that we are able to express prototypically are relations among prototypes; the relations among copies are copies of those relations; and there is no meaningful relation between prototype and copy. We have, after all, just made this clear to ourselves in our analysis of the relation of alikeness. Whereas we know (are familiar with) the copies, the gods know the prototypes. Thus we cannot have true knowledge—namely, the knowledge of prototypes—and the gods cannot have any knowledge of the world as copy. Cornford, too, considers this argument 'almost grossly fallacious'. The Form of the master is not the Master as such, who masters the Slave as such, but Mastership, which is related to Slavery in such a manner that the human master participates in Mastership, and the human slave in Slavery. I find this objection of Corn ford's difficult to understand. The point of the entire theory of Ideas is to clarify what is really meant by abstract terms such as 'mastership', 'slavery', etc., which we employ so matter of course in speaking and writing. The prototype-copy thesis, indeed the hypothesis of separate 'Ideas' ('Forms') altogether, suggests precisely that 'mastership' should be interpreted as "the Master as such"—and against that solution of the problem of the meaning of abstract terms, Parmenides' argument is surely a strong one.

As an aside, let me remark that the attempt to describe divine knowledge can very easily lead us to the conclusion that this knowledge cannot know what we know. For example, logic is the science of true and false judgments and conclusions. But what can a false judgment or conclusion mean in the case of an omniscient being? A being such as this has no need of judgments and conclusions; logic, for this being, would be pointless.

How did Plato conceive of the solution to his problem? Let us proceed on the path of a scientific interpretation in the narrower sense. We translate *physis* as 'nature' in the contemporary sense (assuming we know what we mean by this term), and *paradeigma* as 'example'. Plato himself, in a stylistically striking passage of *The Statesman* (277 d) explicates *paradeigma* as follows[11]:

"The Guest from Elea: It is difficult, oh you who are possessed by a God (*o daimonie*) to demonstrate anything of real importance without the use of an example." (Ernst Kapp used to remark in his lectures, whenever the form of address *o daimonie* occurred: "Now it's becoming uncanny.") "Every one of us is in danger of knowing everything as in a dream, only to wake up and find that he knows nothing.

[11] Weizsäcker's translation (which I follow) hews more closely to the Greek text than do the standard English translations (translator).

Young Socrates: How do you mean that?

The Guest: It does seem odd how I am moving the experience of knowing back and forth in us at the moment.

Young Socrates: How then?

The Guest: I need an example, oh Fortunate One, to show what an example is."

As an example of the essence of 'example' he uses the way in which, with the help of simple, already familiar words, a child practices letters that it cannot yet recognize with assurance in complicated words. The letter in the already familiar word is an example of the Form that it is supposed to recognize in the more complicated word. The word for 'letters' (*stoichea*) is also the philosophical term for 'elements' (LMNts) or atoms. The example chosen is therefore in fact also an example of cognition by means of reduction to the simplest Forms.

Socrates defined Form as an example in nature. Could it be that on principle we know only by virtue of examples? Whatever the resolution of these last difficulties that Parmenides brought up may be, they are in any case connected with the fact that Form and thing are separated from each other like two things. If 'in things' underestimates the separateness of Form and thing, then the reduction of participation to likeness underestimates their intimate association. One cannot identify things on the one hand, and Forms on the other, and then still compare them. What can be known at all, is always *eidos* or, as we have been saying, Form; this is the meaning of the word *eidos*. On the other hand, we know the Form always in examples. Is the Form as known really an example, then? "What is the universal? The individual case. What is the particular? A million cases" (Goethe).

These are chiefly phenomenological assertions, but without them Greek philosophy can hardly be interpreted. They seem, at first, to fit in better with the Aristotelian than with the Platonic *eidos* theory. Don't they say that we know the Form *in* the individual case? Perhaps we can make them clearer by interpreting them in the modern scientific, i.e., cybernetic, manner.

Even animals are able to 'know' Forms, i.e., to react to them correctly; one can in fact say that they do not react to anything else. Correct behaviour can be inborn or unconsciously acquired, or it can be conscious. Human reflecting on conscious insight, the addressing of Form as Form, needs to be further distinguished from this 'comprehending' behaviour. Let us focus on the simplest example, on inborn behaviour. A nestling opens its beak at the sight of its food-bearing mother or of a suitable dummy. It 'knows' the stimulus Form 'feeding' only in examples, but in every and all examples; and it does not distinguish among the examples as far as its reaction 'beak opening' is concerned. Konrad Lorenz once told me of a gander who, as a result of biographical complications and quite against the rule, had two wives, though he always lived with only one at a time. Asked whether the gander was not troubled by this state of affairs, Lorenz answered: "I think he did not know that he had two wives. After all, he was always with only one." This does not mean that the nestling cannot distinguish between the mother and a cardboard dummy, or the gander between the two geese. They are not being distinguished 'as' the Forms 'feeding' and 'wife', respectively, however; both '*are*' the Form.

This is precisely the behaviour that one would expect on cybernetic grounds, if a particular circuit in the central nervous system corresponds to the 'inborn perception of Form'. The circuit responds to a stimulus that can be defined conceptually, i.e., as a Form. If we may call the ability to react in a particular way to a particular stimulus the 'practical concept' of an animal, then the practical concept is itself a Form of behaviour that is lawfully correlated with an encountered Form. The universality of the concept lies in the lawfulness, in the possibility of the repetition of the examples. The practical concept, however, does not 'grasp' the universality, it grasps the universal only in each recurring example. In this sense, the Form is indistinguishable for the practical concept from the example; the Form 'is' the example. It is, to be sure, still a very long way from this point to the theoretical concept, to the cybernetics of reflection that interprets the reflection, too, as a Form of action. I cannot hope to traverse this way in the present essay. Only at a later stage of this way can the Form as such be separated from the example; and perhaps all reflection, at least to the extent that it can be articulated at all, makes use of examples.

On the present level of our considerations it is possible to give meaning also to the concluding difficulty raised by Parmenides; i.e., to the difficulty that the Form of knowledge appears to refer only to the Form of the known, or, in other words, that knowledge (understood as copy) refers only to the known copy. Just as, in the case of the bridle, the rider appears in place of God, so here (for instance) the zoologist. In saying that the zoologist describes the graylag, we have twice used the singular definite article, which designates a Form. Just as the zoologist describes 'the graylag', so the philosopher of the theory of Ideas describes 'the zoologist'. Speaking abstractly, 'the zoologist' describes 'the graylag', but the example of a zoologist describes the examples of graylags he knows. Of course, the individual zoologist is himself already a philosopher to the degree in which he can reflect on his being actually an example of 'the zoologist'; this is what he calls his scientific conscience. Thus, the example of the example is already an example of itself. The simple theories of concept and Form provide the stuff of philosophy, but not its full structure.

In the last sentences we have begun to talk in the language of cybernetics of what Plato and Aristotle call mind (*nous*), and Kant the transcendental subject. We are approaching a more detailed understanding of the question why the prelude of the critique of the Ideas must be followed by the fugue of the philosophy of the One. This, however, is a new and different journey.

References

Picht, G., 1960: "Die Epiphanie der ewigen Gegenwart", in: Höfling, H. (Ed.): *Beiträge zur Philosophie und Wissenschaft. Festschrift für Wilhelm Szilas* (Munich: Francke). Reprinted in: Picht, G.: *Wahrheit, Vernunft, Verantwortung* (Stuttgart: Ernst Klett, 1969).

von Weizsäcker, C. F., 1980: *The Unity of Nature* (New York: Farrar Straus Giroux), pp. 379–400; it was translated by Francis J. Zucker from: *Die Einheit der Natur* (Munich: Hanser, 1971): IV.6 (MD, editor).

Chapter 8
Parmenides and Quantum Theory

Written in 1970 and unpublished until now.[1] According to Platonic Doctrine, the "ideas of things" discussed in the preceding essay[2] must refer back to the higher ideas and, in the end, to the one, the prelude, which deals with a critique of the ideas, is therefore followed by a fugue that criticizes the opinions concerning the One. In modern Science, the organic forms refer back to the universal laws of nature: i.e. in the end to the unity of nature. We encounter this unity in quantum theory. The confrontation of the first hypothesis in the Parmenides Dialogue with Quantum theory is therefore the natural next step. This confrontation is the subject of Sects. 8.4 and 8.5 of this essay, the first three sections being introductory in nature. It turns out that a relationship exists between Bohr's complementarity and Plato's dialectic.

8.1 What Does the Unity of Nature Mean?

We begin by recapitulating the facts and conjectures in which the idea of the unity of nature has presented itself.[3]

The *unity of the law* comes first. This is merely another expression for what physicists call the *universal validity of a fundamental theory*. A 'theory' of this type consists of a number of terms, as well as of fundamental propositions which connect these terms and from which additional propositions can logically be

[1] This text was first published in: *The Unity of Nature* (New York: Farrar Straus Giroux, 1980), pp. 379–400, and is a translation by Francis J. Zucker of: *Die Einheit der Natur* (Munich: Hanser, 1971): IV.6.

[2] See Chap. 7 on "Parmenides and the Graylag Goose" in this volume.

[3] Cf. "Quantum Theory", reprinted as Chap. 7 in: Michael Drieschner (ed.): *Carl Friedrich von Weizsäcker: Major Texts in Physics* (Cham et al.: Springer-Verlag, 2014) and originally published in: *The Unity of Nature* (New York: Farrar Straus Giroux, 1980), pp. 181–222; it was translated by Francis J. Zucker from: *Die Einheit der Natur* (Munich: Hanser, 1971): III.5. Cf. here 7.2: "The Unity of Physics: Part One".

M. Drieschner (ed.), *Carl Friedrich von Weizsäcker: Major Texts in Philosophy*, SpringerBriefs on Pioneers in Science and Practice 23,
DOI: 10.1007/978-3-319-03671-7_8,

deduced. Further, it must be sufficiently clear for practical purposes how the theoretical terms are to be applied in experience, and thus also how the theoretical propositions can be put to the test. A theory has 'validity' only if these procedures are available, and if the propositions thus tested agree with experience. We will not recapitulate the methodological problems implicit in these requirements but will rely for the moment on the fact that, in general, physicists agrees on these matters among themselves. The validity is 'universal' if it extends to all possible objects of a theory; i.e., to all objects covered by the terms of the theory. Here, too, we are satisfied for the moment with practical universality, leaving open the discovery of exceptions or of still more universal laws. We will call a theory 'fundamental' if it extends to all possible objects of nature. The universal validity of a fundamental theory means that all objects of nature are subject to one and the same lawful scheme; it is in this sense that we term this validity the 'unity of the law'. Let me emphasize that all these terms are merely descriptive. They formulate the approximate self-interpretation of contemporary physics, and the following recapitulating reflections will clarify or revise them.

We do have such a fundamental theory—namely, quantum theory. Let us examine in more detail what demands should be imposed on a fundamental theory, and in what sense quantum theory fulfils them.

The theory is to apply to arbitrary objects of nature. To this end, it must be capable of characterizing an arbitrary object. It does so by specifying the totality of its possible ("formally possible") states. The theory must also specify how these states can change in time. These two requirements can be stated from the point of view of classical physics; quantum theory supplements the requirements in its own characteristic way—namely, by fulfilling them.

According to quantum theory, every object possesses, mathematically speaking, the same manifold of possible states[4]; these can be characterized as the one-dimensional subspaces of a Hilbert space. Quantum theory also specifies a universal rule for the composition of two objects into a single object: the Hilbert space of the composite object is the tensor product of the Hilbert spaces of the two part-objects. The theory subdivides the question as to the temporal change of the states into two questions. If the state changes without being observed, it does so in accordance with a unitary transformation of Hilbert space. A particular species of objects (e.g., helium atoms) is characterized by its formally possible unitary transformations, which are mathematically specified by their infinitesimal element, the Hamilton operator H. The Hamilton operator of an isolated object characterizes its internal dynamics and thereby designates certain of its states (for example) as eigenstates of H with particular eigenvalues of the energy. The interaction of the object with other objects is described in terms of the Hamilton operator of the composite object constituted by these objects; this operator can, within certain approximations, be

[4] We will not refer to the Postulate of Finitism (Sect. 7.4. D of "Quantum Theory", cf. footnote 3) in the present description of existing quantum theory. The theory sketched in Sect. 7.5 of "Quantum Theory" (cf. footnote 3) will also not be considered.

reduced to the Hamilton operator of the original object taken as situated alone in a fixed environment. If, on the other hand, the state is observed, then the state changes in another manner. A particular observation admits of only a subset of the formally possible states of the object as possible results of measurement; this subset is constituted by the eigenstates of the Hamilton operator of the object when the instrument of measurement is specified as part of its environment. If ψ was the state prior to the observation, then the probability of finding a particular state φ_n among the manifold of possible results of the observation equals the square of the magnitude of the inner product of the unit vectors in the directions of states ψ and φ_n.

Because of the mathematical formalism that it requires, this description of quantum theory might seem a bit heavy-handed. From the conceptual point of view, the theory may be said to achieve a certain maximum in possible simplicity. The theory characterizes, in unique terms and by means of universally valid prescriptions, arbitrary objects, their composition, changes in their state when not observed, and the prediction of observations. And yet quantum theory, even if we assume it to be universally valid, does not yet express the full unity of nature.

For one can speak, secondly, of a unity of nature in the sense of a unitary character of the species of objects. This character expresses itself in quantum theory in the existence of objects with particular Hamilton operators. Today we believe that all species of objects can in principle be explained as being composed of a small number of species of elementary particles. In the case of inorganic nature, we all believe this to be so; in the case of living organisms, it is the hypothesis on which we have based this book. Finally, we hope to reduce the species of elementary particles to a single basic lawful order, which perhaps we ought not to describe as the existence of a single basic species but rather as the law that specifies all of them.

Thirdly, in the context of contemporary cosmology, it makes sense to talk of the unity of nature as the totality of objects. One speaks of the world as if it were a single object. Quantum theory does indeed permit the composition of arbitrary objects into a new object. It even requires this composition, in the sense that it regards the actual state space of a number of coexisting objects as precisely the state space of the total object they compose; the isolation of individual objects is, in the eyes of quantum theory, always a mere approximation. If the totality of objects in the world can, at least in principle, be enumerated, then quantum theory obliges us in principle to introduce the additional object 'world', which is composed of that totality. At this point, however, certain conceptual problems that form a principal theme of the present essay appear. Let me merely name them for now: If the object 'world' is to exist, for whom is it an object? How are we to conceive of an observation of this object? If, on the other hand, the object 'world' is inadmissible, how are we to describe the coexistence of objects 'in the world' quantum mechanically? Or are we to conclude that quantum theory meets its limits here?

Fourthly, we have tried to base the unity of nature (as conceived under the three preceding aspects) on the unity of experience. We talked, to begin with, of the preconditions of the possibility of experience, and understood 'experience' to already be unified in the sense that 'every' experience may be thought of as connected with every other experience in a contexture of interactions that is free from internal contradictions. This unity appears in Kant under the title of 'the unity of apperception'. In our own approach, which starts not from subjectivity but from temporality, this unity appears as the unity of time. The unity of time (which in our presentation of course embraces space) is, most likely, the only adequate framework for the problem of the totality of objects. With these latter reflections we have delved into the midst of the fundamental problems of classical philosophy. Before confronting these problems, we must still introduce our last approach, the approach of cybernetics.

Fifthly, the unity of man and nature is part of our conception of the unity of nature. Man, in whose experience the unity of nature is discovered, is at the same time part of nature. We try to describe human experience in terms of a cybernetics of truth, which is conceived of as a process in nature. The philosophical problem that arises here is obvious: if this program can be carried through, at least in principle, then the unity of nature is somehow represented within nature as the unity of the experience of man. What does this 'somehow' mean? To put it differently: the subjects, for whom the objects are objects, now form part of the totality of objects. Furthermore, in a cybernetics of truth, human consciousness stands apart from animal subjectivity as a higher-level structure, but the two are also part of a genetic continuum. In the attempt to reduce matter and energy to information, the subjectivity of all substance, if only implicitly and unclearly, is presupposed.[5] The classical formula that nature is spirit which does not know itself as spirit urges itself upon us as a shorthand notation for these problems; but this does not mean that we have understood this formula in the least.

As a next step, we therefore explicitly confront our complex of problems with the ideas of classical philosophy, among which we in fact already find ourselves. Aren't we in the midst of the problems faced by the Eleatic philosopher Parmenides? Hen to pan: One is the totality. The totality is, first of all, the world, "comparable to a well-rounded sphere." But this world embraces experiencing as much as what is experienced, consciousness as well as Being: *To gar auto noein estin te kai einai*, for it is the same to see and to be. I translated noein with "to see" to avoid the abstract introversion of "to think." What can Parmenides teach us?

8.2 A Digression: How Can One Read the Philosophers?

Anyone who turns to the contemporary secondary literature for information on the Eleatic philosopher Parmenides, or on Plato's Parmenides Dialogue, can only fall into despair.

---

[5] Cf. "Matter, Energy, Information", reprinted as Chap. 11 in: Michael Drieschner (ed.): *Carl Friedrich von Weizsäcker: Major Texts in Physics* (Cham et al.: Springer-Verlag, 2014) and originally published in: *The Unity of Nature* (New York: Farrar Straus Giroux, 1980), pp. 274–294; it was translated by Francis J. Zucker from: *Die Einheit der Natur* (Munich: Hanser, 1971): III.5. Cf. here 11.6: "Mind and Form".

How primitive was Parmenides? Was he an astronomical materialist who believed in a spherically shaped universe? Did he suppose—as determined materialists who came after him did—that matter can also think? Was he a pantheist, to whom thinking matter appeared to be God? Was he a spiritualist, for whom the spatial world of appearances was a mere delusion? Is his philosophy the result of his not yet having grasped the difference between consciousness and matter, or between form and matter? Or does the *esti* with the infinitive mean 'one can', so that he would simply be teaching us that reality can be known: "one can think that which can be"? Does he assert that all movement is mere appearance? If so, does he fail to notice that this teaching of his is itself a movement? Is he the victim of a still immature logic? Does he confuse logic and ontology? Is it his reward that he began the search for a rigorous logic? Or was it, rather, that he discovered substance, as the permanent element amidst change? He seems, in any case, to have been a forerunner; but whose forerunner?

And as to Plato's Parmenides: does the 'prelude' relinquish the theory of Ideas, or is it a self-criticism on the road to an improved theory of Ideas, or a preparation for the theory of Ideas? And concerning the first hypothesis of the 'fugue': is it meant 'merely negatively', or merely 'positively', or both? Is it merely a refutation of the Eleatic philosopher Parmenides, which Plato most generously has Parmenides himself deliver? Does it deal with any Idea whatever, insofar as it is an Idea? Does it deal with Plato's One, or with the One of the Neoplatonists? Do these alternatives amount to the same, or are they utterly different? Is the Parmenides a logical exercise, a bit of horseplay, or is it Western theology on its highest level?

The reader is offered all these opinions. Can we hope to learn something that might help us with our own problems from texts that are thus opaque? Shouldn't we turn, rather, to the problems themselves? How can the philosophizing physicist acquire the philological scholarship needed even for merely distinguishing between well-founded and ill-founded textual interpretations?

But it is the problems themselves that urge us to confront these texts. The question as to the nature of physics led us into philosophy, which enquires into the meaning of the terms employed. In Aristotle, Plato, and Parmenides we study these terms at their source; who could inform us more reliably on what these terms stand for than those who coined them? Plato's philosophy itself, however, in inquiring ever more deeply into foundations, ascends from the iron ring to the circle, from the circle to the Idea, from the Idea to the One. It is clear that we have not understood Plato's philosophy so long as we are unable to reconstruct the argumentation in his Parmenides Dialogue. And so long as we are unable to do so, can we hope to understand our own philosophy better than we understand his?

If, keeping this piece of self-advice in mind, we now reread the available interpretations of these two philosophers, we find that every author offers as his explanatory principle the idea beyond which he himself did not progress in his own philosophizing. And since the text never quite fits his interpretations, the deviation from what, in the opinion of the interpreter, the philosopher ought to have said is explained on the basis of the immaturity of his philosophy. As we now turn to these texts, we too, unfortunately, will suffer shipwreck; our interpretations will equally mercilessly reveal the limits of our own philosophizing. In anticipation of these difficulties, let us formulate a few methodological principles for the reading of the texts.

First of all, we should hold to a generalized and simplified version of the principle of the truth of what is being asserted. In the preceding essay ('Parmenides and the Greylag Goose'), this principle was introduced as a philological tool restricted to the interpretation of Plato. It stated that every assertion which Plato puts in the mouth of his characters can be interpreted in such a manner that this assertion is true according to Plato himself. This principle may well be fruitful, though one should not use it indiscriminately; and it applies only to the writings of one philosopher, namely Plato. We are now trying to understand not the assertions of the characters in the Platonic Dialogues, but the opinions of the philosophers themselves. At the points in which we do not understand them, or consider them to be mistaken, our extended heuristic principle now tells us: the philosopher is right. I have not yet understood him if I find myself contradicting him, and I have not yet grasped the truth so long as I do not understand him. Until I have seen the truth he saw, I have no chance at all of seeing a truth that transcends or relatives his truth.

To protect ourselves against merely repeating the doctrines—which also would not mean understanding them—we must keep three points in mind.

First, we cannot expect that discursive thinking can adequately represent what lies at the basis of discursive thinking. In the following section this point will come up again. It means, methodologically, that we must not assume that Parmenides or Plato could have made our task easier by stating their ideas in some other way—'more directly', for example. Nor can we assume that in our own interpretation we could make up for their failure to do so.

Secondly, these philosophers no doubt had to grapple even with the soluble difficulties of discursive thinking and linguistic expression. Once we have understood what they are talking about we are free, indeed obliged, to argue with them. But philosophical texts are by nature ambiguous. To put it in the language of the Platonic theory of Ideas: anyone who is talking of an Idea is automatically also talking of everything that participates in this Idea; anyone who talks of anything automatically talks also of all the Ideas in which this anything participates. This is unavoidable, lying as it does in the nature of meaningful talk. It can therefore happen to us that we have already understood a philosopher on one of the several levels on which he is talking simultaneously (to the extent that such single-level understanding is possible), though we have not yet understood him on some other level.

Thirdly, these philosophers lived at a time other than ours, and are therefore related to us in three ways: they are our teachers, they are our precursors, and they are strangers to us. We philosophize now. Transmitted partly by means of a historical process no longer transparent to us, they are our teachers now. The teacher can point out to us what we ourselves perhaps might not have noticed. In going over his words with whatever understanding we are capable of, more reveals itself to us than we know explicitly. They are our precursors now. Some of the things explicitly available today they foresaw, some they did not foresee even though these things developed from their beginnings. In repeating their words, we are justified in recognizing potentialities that could not have been explicitly visible to them. They are strangers to us now. Their cultural environment has disappeared, and we know that every man remains a stranger to all other men, even to his contemporary and to his friend, both in matters of which he is aware and in those of which he is not. Therefore, every successful interpretation is at the same time a productive misunderstanding. That, too, is part of human existence in history, is part of the *condition humaine*.

Keeping all this in mind, let us, in the naiveté of the question as to the nature of truth, venture a dialogue with the philosophers.

8.3 What Were Parmenides and Plato Talking About?

Our topic is not the multifariousness of Platonic politics and ethics, physics and logic, nor the multiformity of the universe of Ideas. Rather, our topic is the theme which Plato shared with Parmenides and which he himself presents in the Parmenides Dialogue under the title of the One. The topic deals with the unity of Being, with the Being of the One, with the unity of the One.

Let us begin with Plato, who seems to have come down to us, through his writings, in his entirety—the first philosopher to have done so. But the explicit text of his writings almost fails us: except for the Parmenides, none of his works discusses the role of the One in his system of thought, and what the Parmenides offers is a total paradox. The connection with the Idea of the Good in The Republic was suggested by Aristotle[6]; the correct connection with the highest species of Ideas in The Sophist is largely unknown. We know no more about it than Plotinus did, probably less. For surely in Plotinus' time a more comprehensive written tradition was still extant, perhaps also a trustworthy oral tradition.

We therefore direct our attention to the question of Plato's unwritten doctrine. Indeed, all his Dialogues border on the unwritten: they challenge us to think beyond them. A Dialogue frequently ends in a paradox, and a later Dialogue solves this paradox only to end in a paradox on a higher level. By annotating every Platonic text with the parallels in other Dialogues, we obtain a system of well-matched hooks and eyelets, a contexture that reveals more than does a cursory reading of the texts. Even the assertion that Platonic philosophy deals with the ascent from the iron ring or ball to the One is still a rather naive attempt at thinking beyond the Dialogues. Aristotle tells us that Plato taught doctrines that he never wrote down (the *agrapha dogmata*).[7] Could these help us along?

Let us first ask why Plato should have chosen not to write down certain of his doctrines. Either he thought it impossible to do so, or he thought it possible but

[6] *Metaphysics* 1091 b 13, *Eudemian Ethics* 1218a19f.

[7] Cf. Kramer (1959); Gaiser (1963).

undesirable, or perhaps he thought it desirable but did not get around to it. The core of the doctrine of the One was probably of the first sort; what Aristotle transmits to us might be more of the second, in peripheral areas also of the third sort. But why was Plato of the opinion that certain doctrines could, but had better not, be written down? What he says in the Phaedrus and in his Seventh Letter suggests that these doctrines are intimately connected with what cannot be written down at all; anyone who has not understood this point can only do harm with the doctrines. According to the testimony of Aristotle, the unwritten doctrine of the second sort appears to have been a two-principle metaphysics, and a mathematical natural science developed from such a metaphysics. This doctrine seems to have been a descending construction of what the soul, in its stepwise ascent to higher knowledge, recognizes as the diverse levels of Ideas—a notion criticized in the prelude of the Parmenides. The two principles are the One (hen) and the unlimited duality (aoristos dyas). Their interplay gives rise to the numbers, to the spatial dimensions and figures, and to the elements in the world of perception. What can we learn from this doctrine, if we have no scholarly interest in the games played by the hypothetical-speculative natural science of the ancients, but wish to pursue questions valid for us, too?

The two-principle doctrine contains a fundamental paradox. Principle means beginning (arche). A multiplicity can be analysed with respect to any number of relative principles. This is what Aristotle repeatedly does with his phenomenological method.[8] But if I am not mistaken, Aristotle escapes from the real, speculative problem posed by the multiplicity of principles by means of the pros-hen structure of his doctrine of categories, and by means of the doctrine of God as the highest ousia. It is this concealment of the ontological difference that Heidegger refers to as metaphysics. Let us first consider a rather naive formulation of the speculative problem of the two principles. We note that several, or even only two, 'beginnings' are no beginnings at all, since we still have to face the questions: Why these two? What do they have in common (e.g., being a 'beginning')? In what do they differ? If there is to be a beginning, it had better be a single one. But how can a beginning be One, if its consequence is (if it explains, if it evolves into) a plurality? If the beginning is to be One, then nothing must exist beside it. It must be all: One is the whole. We have arrived at Parmenides of Elea. Or have we really?

We postulated that something like a beginning might exist, and drew our conclusions discursively. We argued with 'ifs' and 'buts'. (In conversation, Heidegger referred to this procedure as 'arguing around'.) We drew a conclusion that, if correct, denies all we started with. The discursively correct consequence would be that we admit to having performed a *reductio ad absurdum*: there can be no Parmenidean One; neither, therefore, can there be a single unique principle, and thus in the strict sense of the term there can be no principle. We have not arrived at Parmenides.

---

[8] W. Wieland: *Die aristotelische Physik* (Göttingen: Vandenhoeck und Ruprecht, 1962). Though overemphatic in its language-analytical interpretation, this work offers an excellent analysis of the pluralism of principles.

If we can trust in the truth of what he asserts, Parmenides acted quite differently.[9] He begins his poem, which strikes one as highly abstract, with the image of his ecstatic transport to the gates of wisdom, which open on the goddess of truth. Look! she bids him; he sees. Introduced in the traditional language of the mysteries, the poem is the epiphany, the manifest appearance of that which is. That which is, to eon, is the One that he sees. And he must also learn that everything else is not, that it is merely the opinions of human beings. In Picht's words, we understand that which is as the eternal present, for which we have been historically prepared by the doctrine of the presence of the divine nous—the divine seeing—with all things, those that are, those that were, and those that shall be.

I will not try now to develop the content of the Parmenidean doctrine, and refer the reader instead to Picht's interpretation. What matters to us at this point is the question of how we ourselves react to the possibility of this kind of knowledge. It combines in an inseparable whole the assertoric form of a report on what has been seen directly—Picht says, pointedly: the poem is itself the epiphany—with the most extreme abstract rationality of arguments and assertions. Can these two really go together, or do they not rather constitute an internal contradiction that explains the confusion of the interpreters? How can the appearance of the divine be combined with scientific rationality?

Let us, for the purpose of comparison, return to the everyday of science. In our description of physics we saw[10] that physics is based on universal propositions that experience can neither verify in their universality nor, in a strictly logical sense, even falsify. We talked of 'scientific perception' as a kind of perception of structure. This structure we identified hypothetically with the Platonic Form (Idea) in the individual thing,[11] and from the cybernetic point of view this entity did seem conceivable to us.[12] The basic material of scientific knowledge is available to us in a perception of structure that carries with it no sense of illumination, precisely because of its so common availability. But the great, the new steps in science are based on such perceptions, too, perceptions now of previously concealed structures, which we said were characterized by simplicity, universality, and abstractness.[13]

We need to make clear to ourselves the methodological role of scientific perception. It may be simplest to exemplify this role in the unusual case of a great theoretical step forward. The scientist who formed the new idea experienced

---

[9] I essentially follow the interpretation of G. Picht, "Die Epiphanie der ewigen Gegenwart" in: *Beiträge zur Philosophie und Wissenschaft. Festschrift für Wilhelm Szilasi* (Munich, 1960). Reprinted in G. Picht, *Wahrheit, Vernunft, Verantwortung* (Stuttgart: Ernst Klett, 1969).

[10] Cf. Sects. 3.4.4–3.4.5 in this volume; cf. also I.5 of C. F. v. Weizsäcker: The Unity of Nature (New York: Farrar Straus Giroux, 1980), pp. 72–83; it was translated by Francis J. Zucker from: Die Einheit der Natur (Munich: Hanser, 1971): I.5.

[11] Cf. "Parmenides and the Graylag Goose", Chap. 7 of this volume; here Sect. 7.4.

[12] Cf. "Models of Health and Illness, Good and Evil, Truth and Falseness", Chap. 6 of this volume; here Sect. 6.5.

[13] Cf. Sect. 3.4.5 in this volume.

something like an illumination; he saw what others and what he himself had previously not seen. But he must not invoke his illumination in support of the idea, neither before others nor before himself. He must ascertain whether he has seen truly by drawing inferences from his new idea and testing his inferences with respect to already available or newly produced experiences. He is duty bound to try to falsify his discovery. If it is a true discovery, it will resist all falsification attempts and will render comprehensible what was previously incomprehensible. The discovery justifies itself as does a light kindled in the dark: by means of what it teaches us to see. The scientist will convince others if he can make them see what he sees. And the experiences which are required for falsification, or through which new knowledge becomes possible or comprehensible, are all of the nature of the perception of structure—normally of the undramatic sort, involving structures that have long since been known. Every single so-called experience must on principle be submitted to this kind of criticism; we must be able to check it, as it were, and checking means seeing the same thing over and over again, and also seeing its consequences.

But this is precisely the methodological structure of Parmenides' poem. In poetic language—i.e., in the language familiar to the members of his culture—Parmenides describes that he was led to see, he explains what he sees, and he presents arguments that a trained mind must find convincing; in this way he teaches the reader to see for himself. If we fail to see, the fault perhaps lies only in our own shortcomings. But if Plato, clearly talking of the same subject as Parmenides, nevertheless criticizes him,[14] it must be possible to disagree about this perception. The disagreement is not about the fact that something has been perceived, but about our understanding of what has been perceived; its resolution requires further perceptions.

Although sense perception, too, has a predicative character and perceives structures that can show up in argumentation, it is not experienced as an act of discursive thinking, it is not part of a context of arguments. Human tradition is acquainted with an experience that is related to what has here been presented in the form of arguments just as sense perception is to its virtual conceptual content; in the West, we call it the mystical experience. The mystical experience is culturally conditioned in its manner of expression, yet at the same time, and to a surprising degree, it is identical in all cultures. Its highest concept is that of 'becoming one', the *unio mystica*. Becoming one can mean that two merge into one. It can also be understood as 'becoming the One' (in the sense of: becoming an adult, or a beauty). In the Neoplatonic school, the One of mystical experience was equated with Plato's One. In the old Asiatic traditions, meditative training is considered an obvious precondition of philosophical thinking that, on its higher levels, interprets the higher levels of the meditative experience.

On these levels the question of whether the One is to be represented by God reverses its significance. Religion has acquainted us with the idea of gods or of a God. This idea, as an idea, is familiar even to the irreligious. It is one of the

[14] Explicitly so, for instance, in the *Sophist* 241 d 5.

elucidating ideas—like matter, consciousness, universe, and love—by means of which we try to gain access to the abstract concept of the One, which is not available to us in experience. Is the One perhaps an abstract designation for one of these familiar realities or ideas? Both philosophy and meditative experience must reverse the direction of this question. What these ideas mean is now the question, and the answer lies in a recourse to the One. If we call the One God, then God is a name for the One. The actual universe, with all its matter, its consciousness, its loving and desiring, is then an image of the gods (Plato's agalma, Timaeus 37 c 8) or the work of God (also in the Timaeus; this is how Christian theology later interpreted Genesis 1); the gods of the universe are appearances or derivatives of this God. In the poem of Parmenides, seeing and Being—or, as we would put it, awareness and Being—are united in what is one and the same; or better (following Picht), the identity (the same: tauton) is; i.e., it lets be both the seeing and the Being. In the Indian Vedanta doctrine, the One is sat-chit-ananda, which we can translate as being-awareness-bliss. T. M. P. Mahadevan once explained the Advaita (non-duality) doctrine to me by saying that these three are not aspects of the One but are identical with it, and they separate out only in the domain of temporal appearances; sat is in all there is, chit in every awareness, ananda (bliss) only in a purified awareness.

The recognition of a meditative or mystical experience of unity is not an escape from rationality but, assuming we have argued correctly, a consequence of the understanding of the nature of rationality. Discursive philosophy can be viewed as a preparation for, or an interpretation of, this experience; or it can also be an interpretation of the recognition of the possibility of this experience. The mystics have in fact found the philosophy of the One to be an interpretation of their experience. And it seems obvious that, conversely, those who reject the possibility of this experience or consider it irrelevant would find the philosophy of the One to be incomprehensible or confusing, and would find a way of escaping from this state into some oversimplifying interpretation. On the other hand, mystical experience itself is as remote from being philosophy as sense perception is from science. An essay such as this, which seeks to discuss these matters in the medium of our contemporary scientific awareness, can at best offer philosophy as an interpretation of the recognition of the possibility of mystical experience. It must try to argue theoretically concerning the One. Plato, too, made this effort in his written doctrine, and in particular in his Parmenides Dialogue.

8.4 The First Hypothesis of the Platonic Parmenides and Quantum Theory

We return to the unity of nature as it presented itself to us in the five introductory, recapitulating reflections. We ask what Parmenides and Plato can teach us on this subject. If the Parmenides of the Platonic Dialogue was right in thinking that he was presenting an exercise (gymnasia) necessary for the understanding of Forms

(Ideas),[15] then this exercise will do us good, too. This undertaking will be limited in two respects: we engage in Plato's gymnastics only with a view to the current state of science, without trying to interpret Plato's philosophy; and we juxtapose physics only with Parmenides and Plato, leaving aside Christian theology, the philosophy of subjectivity, and the unity of historical time in the contemporary philosophical sense. We do an exercise in reflection, no more than that. This accounts for the odd confrontation in the title of the essay; surprisingly enough, the confrontation does seem to produce some results.

We can begin at 137 a 4 with the preparation for the first hypothesis. Parmenides asks: "Where shall we begin, then? What supposition shall we start with (hypothesometa)? Would you like me, since we are committed to play out this laborious game, to begin with myself and my own original supposition? Shall we make our hypotheses concerning the One and consider the consequences of assuming that it is one or that it is not one?"[16] Here the translator faces his first difficulty: in the part of the sentence reading *peri tu henos autu hypothemenos, eite hen estin eite me hen* one can just as well interpret the eite hen estin as an independent expression ('that the One is'). Also, one can read the first part of the sentence at the actual beginning of the hypothesis, 137 c 4, the 'theme of the fugue', *ei hen estin*, either as the independent 'if the One is' or, as before, as 'if it is one'. Some interpreters therefore hold the first hypothesis to assert that the One is, others that the One is one.[17] Of course the dilemma might be that of the traveller at a fork in the road where there is no signpost: perhaps both roads lead to the same destination and are unmarked for that reason. For all interpreters agree that the *ei hen estin* of the first hypothesis accents the hen, in contrast to the hen ei estin of the second hypothesis; the first hypothesis is therefore concerned with the unity of the One, the second with the Being of the One. If the first hypothesis is true, i.e., if the One is one in a strict sense, then the two grammatical constructions appear to be saying the same thing.

But what is the One that is here being discussed? Our preceding reflections prompt us not to expect we might be able to point to something familiar and say: that is what's meant. It is surely certain that nothing of the sort can be meant. Nevertheless, the One must somehow be (perhaps always has been) familiar to us, for how could Plato otherwise enable his Parmenides and Aristotle to conduct a smoothly flowing discussion on it, in which each apparently understands what the other is saying? The argument reads as if it should be understandable by itself. What common knowledge does it presuppose? This knowledge is threefold, it seems to me: First, since Parmenides expressly points to himself and his hypothesis, we ought to be familiar with his didactic poem and make use of it. Secondly, since he argues by means of familiar conceptual meanings, we should try to grasp his concepts so as to be able to understand, or at least to follow, the

[15] Cf. footnote 4 of "Parmenides and the Greylag Goose" in Chap. 7 in this volume.—*Translator.*

[16] In the last sentence of this quote, Weizsäcker's translation differs from Cornford's; I follow the former. (Cf. the preceding footnote.) —*Translator.*

[17] Thus especially Suhr (1969).

arguments. Thirdly, the whole Platonic philosophy, which Plato assumed the reader to be familiar with, must be seen as standing in the background. Not that the rules of the game allow Plato's philosophy to be used as an argument, but its recollection is a permissible aid in interpretation. More specifically: First, our topic is the eon, which Parmenides himself designated as one. The arguments of the first hypothesis show, secondly, that this eon cannot be characterized as Parmenides had tried; in this sense the first hypothesis is certainly a critique of the Eleatics. Thirdly, the One in its strict unity is thereby assigned a particular place in Platonic philosophy, a place for which we must search.

The arguments actually make use only of what has been presupposed explicitly: the unity of the One and the meanings of terms that must have been familiar to the philosophically trained reader in Plato's time. Although the terms being used are taken from the list of the Parmenidean 'characteristics' (*semata*),[18] they appear also in the Aristotelian categories; we may therefore take them to be basic concepts that were well established since the Eleatics. But then the argument must be intended as stringent on the basis of these presuppositions alone. If at the same time it also corrects the doctrine of the old Parmenides, it can do so because this doctrine was based on the same premises. We may therefore say, in agreement with Lynch,[19] that the first hypothesis refers to everything that is one. The argument serves well as a critique of the Eleatics (contrary to Suhr) precisely for the reason that it is valid philosophy and covers everything that is one in Plato. Of course the question is now what 'one' designates in this sense. We lose our red thread entirely if we turn to a doxology of Platonic doctrines and read: every Idea is intended, for it is one; or: the familiar One of Plato is intended. The problem is, after all, to discover what 'an Idea' or 'the One' means in the first place.

We therefore test the stringency of Platonic argumentation by applying it to quantum theory.

> Well then, said Parmenides, if there is a One, of course the One will not be many. Consequently it cannot have any parts or be a whole. For a part is a part of a whole; and a whole means that from which no part is missing; so, whether you speak of it as 'a whole' or as 'having parts', in either case the One would consist of parts and in that way be many and not one. But it is to be one and not many. Therefore, if the One is to be one, it will not be a whole nor have parts (137 c 4–d 3).

In classical physics, no such One exists, with the possible exception of the point mass. In quantum theory, not even elementary particles are point masses; rather, they contain other elementary particles virtually and exhibit spatial extension in certain experiments. Instead of elementary particles, we had therefore better consider either the arbitrary individual object (for it is one object) or, more specifically, the universe. According to classical physics, the universe is composed of many objects, thus perhaps constitutes a whole, but surely not a One in the strict sense. What is the situation in quantum theory?

---

[18] M. Suhr: *op. cit*, pp. 25–31.

[19] See Lynch (1959).

We know the rules for the composition of part-objects into a composite object. Should we regard all quantum theoretical objects as composed of parts, or are some of them thus composed while others are not? We must first criticize the term 'composed', and distinguish it from 'divisible'. It is frequently, and correctly, said that according to quantum theory, the hydrogen atom (for example) forms a unity that is destroyed by localizing parts in it; i.e., the nucleus and the electron. The atom is sometimes referred to as a whole, but in a sense that differs from Plato's: one doesn't mean that no part is missing, but rather that the parts are 'submerged' in the whole. At any rate, we can, if we like, adapt quantum theoretical to Platonic terminology by calling the quantum theoretical object a One.

This terminology turns out to be rigorously correct if we regard it as an expression of the mathematical form of the composition rule. Among all possible states of a composite object there exists only a set of measure zero in which its part-objects are in specifiable states; only in these 'product states' is it rigorously correct to assert that the part-objects exist.[20] In all other states one can assert only: if the composite object is subjected to a measurement that forces the appearance of part-objects, then they will show up with such and such probability in such and such states. The 'composite' object is therefore a One which can be divided into many, but which then ceases to be what it was. The special case of the universe will be considered below.

Plato takes up the spatial characteristics next. The One has neither beginning, middle, nor end, it has no form, being neither straight nor round. It is located nowhere, neither in another nor in itself. It neither rests nor moves. For all this would be possible only if it had parts (137 d 4–139 b 3). I do not wish to discuss Plato's arguments in detail, but rather ask what quantum theory has to say.

To be able to state that an object has a particular (contingent) property—for instance, that a particular observable X has a particular value ξ—either X must be measured with the result ξ, or a state must exist in which the probability of finding the value ξ if X is measured is one. The set of object states in which a given observable has any particular values whatever is again of measure zero. Furthermore, as is well known, there are no states at all in which an object can have fully determinate values of both position and motion; that is what the Principle of Indeterminacy is about. Thus, in itself, a quantum theoretical object is one and does not simultaneously have both a determinate position and a determinate motion. The next step is to note that the spatial properties of objects can be ascertained only by means of interaction with other objects. An interaction described purely quantum mechanically takes the form of an internal dynamics of a composite object consisting of all the interacting objects; the original object has become 'submerged' in this totality. The original object itself is measured only

---

[20] In Drieschner's axiomatic treatment, this point turns out to have fundamental significance. Cf. "Quantum Theory", reprinted as Chap. 7 in: Michael Drieschner (ed.): *Carl Friedrich von Weizsäcker: Major Texts in Physics* (Cham et al.: Springer-Verlag, 2014) and originally published in: *The Unity of Nature* (New York: Farrar Straus Giroux, 1980), pp. 181–222; it was translated by Francis J. Zucker from: *Die Einheit der Natur* (Munich: Hanser, 1971): II.5.

when an irreversible event occurs in the objects with which it interacts, and which we call the instrument of measurement. But irreversibility is not part of the quantum theoretical description of a state; it designates, rather, the transition to a classical description, to the description of the knowledge of finite things by finite beings. As a result of this transition, a part of the quantum theoretically possible information on the total system (the phase relations between object and measurement device), and thereby the unity of the total system, is necessarily sacrificed. We may therefore say: spatial determinations become possible only when quantum theoretical unity is lost.

We can now apply what we have learned to the universe. Actually, it is never legitimate to describe any object in the world whatever as an isolated One. No object would be an object in the world were it not for the interaction between the two. But then, strictly speaking, the object is an object no longer. Were there such a thing as a quantum theoretical object in the strict sense of the term, then it would have to be the whole universe. If we now apply to the universe what we have learned about objects, we must say: the description of the universe as a spatially structured whole—i.e., as a collection of separate parts embedded in space—stands in an exclusive relation to its description as a quantum theoretical unity. This is true even though, viewed mathematically, the quantum theoretical description, which includes the phase relations, is richer, not poorer than the spatial description. If the quantum theoretical description is extended to the entire universe, however, then nobody is left to know of this information. The One as such cannot even be known potentially. But that is Plato's conclusion, too: "Consequently it cannot have a name or be spoken of, nor can there be any knowledge or perception or opinion of it" (142 a 3–4). To put it quantum theoretically: the larger the object of our knowledge, the more knowledge can be gained about it of the sort that cannot be described spatially. If we incorporate absolutely everything into the object, thus including also our own knowledge of it, then this knowledge becomes fictive, i.e., only formally possible, and thus no longer satisfies the preconditions for being known. Perhaps this fiction is the shadow cast by a non-finite, divine omniscience on the wall on which we record our finite knowledge; finite knowledge, however, will not redeem this claim.

It should be stressed, of course, that this entire argument ignores the temporality of our knowledge. But the fundamental concepts of quantum theory are temporal. Unity is mediated by the phase relations, and these signify probabilities, thus future possibilities. Between the unity of the many in nature and the unity of the One we discover the unity of time. Since this transcends the Platonic approach, it will not be discussed in this essay.

In the foregoing we arrived at the Platonic conclusion while skipping the middle part of the argument. Plato shows (139 b 4–140 d 8) that the paired terms identity-difference, likeness-unlikeness, and equality-inequality cannot be applied to the One either. The One can neither be identical with, nor different from another, etc. The essential point is that the definition of the One coincides with none of these other definitions. It is a highly interesting question what logic Plato used in this argumentation, and whether his inferences are stringent or—as it must seem, in the light of some of the interpretations—contain logical errors. Again, instead of dealing with this problem in the interpretation of Plato, we apply his reflections to quantum theory. As in the case of spatial determinations, we must also operationalize the categorical determinations just mentioned in order to apply them to objects. This means interaction, and thus the loss of the unity of the object. To find out, for example, whether an object X is tauton with an object Y in the sense of the eidos—i.e., whether it is of the same species—one must observe the behaviour of both. This holds also if the assertion that the object is of the same species as itself is not merely intended as an empty formula but as subject to empirical testing. Even the numerical identity of an object with itself requires observation: the non-classical symmetries that lead to the Einstein-Bose and the Fermi-Dirac statistics derive precisely from the fact that one cannot pin down the numerical identity of an object with itself. If we are to hold on to an object as a One in the strict sense of the term, it would have to be completely isolated; but then even its identity with itself becomes unobservable.

As a final group of concepts prior to the conclusion, Plato discusses the concepts of temporality (140 e 1–141 e 7). Neither earlier nor later ('older' nor 'younger') is applicable to the One. The One neither was nor will be nor is now. On applying this to quantum theory, we are made aware of an inconsistency in the usual presentation of that theory. The quantities that characterize an object (the state vector in the Schrödinger, the operators in the Heisenberg picture) are written as functions of a parameter t, which is identified with time. Although considered measurable in principle, time is the only measurable quantity to which no operator corresponds. Some other observable is always measured in the place of time—an observable whose variation in time is theoretically sufficiently well known (and, preferably, periodic). The isolation of an object one wishes to hold on to as a unity of course nullifies the measurement interactions required for the determination of its states in time. A strictly isolated object does not exist in time either. This of course nullifies the meaning of the fundamental concepts of quantum theory, and in particular that of the probability concept—of all the concepts, in other words, with which we formally describe an isolated object.

Plato accomplishes the transition to the concluding paradox in a manner that must seem startling from the point of view of the average interpreter. We are usually told that, for Plato, true Being is found in the ideas, and that their Being is non-temporal. Let us note his quite different emphasis in the following passage (141 e 3–142 a l)[21]: "If therefore the One does not participate in any way in any time, then it never has become, nor was becoming, nor was it ever; neither did it become now, nor is it becoming, nor is it; and neither will it become later, nor will it have become, nor will it be.—This is as clear as can be (alethestata).—Can a thing participate in Being (ousia) in any other way?—It cannot.—In no way, therefore, can the One participate in Being.—It appears that it cannot.—Therefore, the One in no sense is.—So it seems.—It cannot even be in the sense of being one, for then it would already be a thing and participate in Being. But, if we can trust this manner of arguing, it appears that the One neither is one nor is it." This is

[21] Here I follow Weizsäcker's translation throughout, not Cornford's.—*Translator.*

followed by the passage, from which we have already quoted, that there can be no knowledge or even mere opinion of the One. "Is it really possible that it is thus with the One?—I do not think so" (142 a 6–7).

The point is that Being can only be in time. Is the partner in dialogue being consciously misled? I doubt it. One will have to distinguish the time which abides in the One (aion in Timaeus 37 d 5) from its copy (chronos) which progresses in accordance with number and is counted by the celestial movements.[22] But we cannot pursue Plato's ascent any further here.

Is the final paradox a refutation of the hypothesis? Who can help thinking in this connection of the passage in The Republic (509 b 9) in which the Idea of the Good is said to lie beyond Being (epekeina tes ousias)? Indeed the text does lead to an explicit contradiction: if the One is one (137 c 4), then the One cannot even be in the sense of being one (141 e 10–11). Now the prohibition of self-contradiction is a characteristic of Being: what contradicts itself cannot exist, it cannot even be meaningfully asserted. To wish to assert something that 'is' beyond Being would indeed be an absurdity. The theologians who claim that the One, in far surpassing the domain of all Being in dignity and power (The Republic 509 b 9), is not in need of our saying so, are as justified in appealing to this passage as are the logicians who assure us that it deals with nothing at all, and that, consequently, nothing can be said about it. Both take Plato at his word.

A decision can be made only if we see whether another road exists that is more satisfactory to logicians, or whether this very self-contradiction is actually required for there being such a thing as a domain without self-contradiction. The decision is made after the further hypotheses have been worked through.

8.5 The Approach to the Second Hypothesis

In the present context, we cannot concern ourselves with the scope of Plato's philosophy as sketched out in the further hypotheses. But we must at least discuss the approach and its foremost consequence.

If the One is, then its unity must be distinguished from its Being. But then it is already twofold: it is one and it is. In turn, both of these are twofold: the One, in that it is; and what is, in that it is one. This procedure is to be iterated ad infinitum. The One, if it is, contains an infinite multitude (142 b 1–143 a 3). Among this multitude, numbers as well as other high conceptual species are demonstrated to exist (such as 'difference', by means of the difference between unity and Being). The One, in being, unfolds into the world. This world is, to be sure, full of inescapable contradictions that were already included in the beginning. "Thus not only is the One, in being, many, but also the One itself is distributed by Being and is necessarily many" (144 e 5–7). The logician will not escape self-contradiction

---

[22] On this point and on the entire chapter, cf. Wyller (1960).

even in the world that is. He can, to put it crudely, fixate a given One, a One that is, and describe it without internal contradiction, but only so long as he does not inquire after its origin or its subsequent apportionment; i.e., so long as he does not inquire after the way in which its unity can be, or in which its Being can be unitary.

Let us turn once more to quantum theory. The way in which an object, which at first is conceived of as completely isolated, can nevertheless be an object, i.e., can actually be, is through its interaction with other objects. But in the same measure it ceases to be precisely that object, indeed to be an (one) object at all. One can say, paradoxically: any property of an object becomes observable only by the object's losing that property. The approximation within which this loss can be discounted is classical physics, or the classical ontology on which classical physics rests. But it is only within the classical approximation that we can make and formulate our observations. In this sense, all physics rests on an approximation. In a given case, this approximation can itself be described in terms of physics and thereby made ever closer, but only by again invoking it in another place.

Bohr has described this situation by means of his concept of complementarity. This concept has been widely viewed as resignation in the face of incomprehensible empirical difficulties in the measurement process, its application by Bohr in other areas therefore as the illegitimate hypothetical generalization of a problem in physics. We have now discovered, however, that complementarity already announces itself in the Platonic Parmenides. It is in fact classical ontology that falls short of the level on which Parmenides (the old Eleatic as well as the Platonic Parmenides) reflected; classical ontology fails to realize that its application presupposes its falseness. The universe can be only insofar as it is not one but many. But contrary to the description offered by logic and classical ontology, the many do not exist in themselves; they exist only in the One, which is beyond thought.

In conclusion, we cast a final glance at the doctrine of the two principles. Two principles, we said, are no principles; what they have in common and what distinguishes them would be their principles, and these would again be two. One principle, on the other hand, does not lead to the many. The first two hypotheses explicate this problem; they show that it could not be otherwise. Plato's two principles, as transmitted by Aristotle, designate (in specialized terminology) unity and multiplicity. Unity by itself is not a principle; in being, unity is multiplicity, but at the cost of self-contradiction.

So much, then, on Plato and quantum theory.

References

Gaiser, K., 1963: *Platons ungeschriebene Lehre* (Stuttgart: Ernst Klett).

Kramer, H. J., 1959: *Arete bei Platon und Aristoteles* (Heidelberg: C. Winter).

Lynch, W. F., 1959: *An Approach to the Metaphysics of Plato Through the Parmenides* (Washington, D.C.: Georgetown University Press).

Suhr, M., 1969: *Platons Kritik an den Eleaten* (Hamburg: Stiftung Europa).

Wyller, E. A., 1960: *Platons Parmenides* (Oslo: Norske Videnskaps).

Chapter 9
Possibility and Movement: A Note on Aristotelian Physics

Published in the Festschrift for Josef Klein on the occasion of his 70th birthday (Göttingen, 1967). This essay deals with the potential conception of the continuum, which I tried to sketch out as a problem of physics (i.e., rather than pure mathematics) in an essay entitled "Kontinuität und Möglichkeit" [Continuity and Possibility], written in 1951 (reprinted in *Zum Weltbild der Physik* [Stuttgart: Hirzel, 1958][1])—The consistent treatment of this conception requires a theory of time that not even my "Quantum Theory"[2] has been able to offer.

With his conception of the infinite as potential, developed in Books III and VI of his Physics, Aristotle accomplished enough for two millennia—until the time of Kant and Gauss.[3] Indeed the contemporary reader can no longer understand Kant's statements on the infinite and on continua unless he realizes that on this point Kant presupposes, as the only conceptually rigorous way of speaking about the infinite, the Aristotelian conception—for example, when he frequently stresses that the parts of space are 'spaces', or when he denies (in the *Metaphysical Foundations of Natural Science*, Chap. 3) to infinite absolute space, in contrast to the empirically real, relative spaces defined by moving bodies and always conceived of as finite, the character of an 'object'. And from Gauss, the great preceptor of mathematical rigor, we have learned the explicit requirement to conceive of the infinite as

[1] The English translation of this book: *The World View of Physics* (Chicago: University of Chicago, 1952), does not contain this essay.—*Translator.*

[2] Reprinted as Chap. 7 in: Michael Drieschner (ed.): *Carl Friedrich von Weizsðcker: Major Texts in Physics* (Cham et al.: Springer-Verlag, 2014) and originally published in: *The Unity of Nature* (New York: Farrar Straus Giroux, 1980), pp. 181–222; it was translated by Francis J. Zucker from: *Die Einheit der Natur* (Munich: Hanser, 1971): III.5.

[3] This text was first published in: The Unity of Nature (New York: Farrar Straus Giroux. 1980), pp. 346–356, and is a translation by Francis J. Zucker of: Die Einheit der Natur (Munich: Hanser, 1971): IV.4.

M. Drieschner (ed.), *Carl Friedrich von Weizsäcker: Major Texts in Philosophy*, SpringerBriefs on Pioneers in Science and Practice 23,
DOI: 10.1007/978-3-319-03671-7_9,

merely potential. Neither Kant nor Gauss reflected on this conception historically, as a conception we owe to Aristotle; in treating it as so self-evident that it no longer had to be linked to his name, they paid Aristotle the greatest homage a thinker can gain. The potential conception in fact avoids the paradoxes that critical minds have discovered again and again in the conception of an actually exist-ing infinite. How it does this is made especially clear in the brilliant presentation of the Aristotelian theory of the continuum due to W. Wieland.[4] In intention and in its largely successful execution, this theory is phenomenology pure and simple; avoiding constructions as much as possible, it is a description of the way in which the infinite and the continuum are actually given to us in the interplay of perception and thinking (αἴσθησις, and νοῦς).[5]

Anyone who has understood this may wonder why mathematics nevertheless shifted in the second half of the nineteenth century to the conception of the infinite as actual, which is so much more difficult to justify. So successful has this shift been that it is nearly impossible to disabuse the contemporary student of mathematics of the superstition that this conception is the only possible, indeed 'the' theory of 'the' infinite and of 'the' continuum. The reason can be found in internal mathematical problems. The rigorous foundation of the infinitesimal calculus as intended by Weierstrass and Dedekind seemed to require a theory of irrational numbers, which in turn appeared to be impossible without Cantor's concept of actually infinite sets. The discovery of the set-theoretical paradoxes, however, destroyed Cantor's idea of the simple givenness of the basic set-theoretical concepts, as well as Frege's attempt to provide these concepts, whose questionable nature he clearly perceived, with a purely logical derivation. The shift to a purely axiomatic set theory meant relinquishing the self-evidence of the basic terms; henceforth, set theory could be justified solely on the grounds of its beauty and usefulness. And yet one cannot prove the self-consistency of an axiomatic system without using formal reasoning that is acknowledged to be self-evident. Mathematicians like Brouwer, Weyl, and Lorenzen, who try to construct a theory of the continuum taking this self-evidence as their point of departure, re-turn to the potential conception, which they render more precise by means of concepts such as 'constructive' and 'operative'.

The intuitionist way of thinking is most likely the only way in which to make the contemporary mathematician receptive once again to the ideas of Kant and Aristotle. For these philosophers do not ask what can at all be thought up ('axiomatically', in the contemporary sense), but what has 'always already' been understood when terms like 'limit-less' (ἄπειρος) or 'connected together'

[4] See Wieland (1962).

[5] I have to leave aside in this Note the question as to what Aristotle owed to older philosophical (Pythagorean, Eleatic, Platonic) or mathematical (Pythagorean, Eudoxian) conceptions, or in which points he may even have fallen behind them.

(συνεχής) are rationally employed. Wieland shows, however, that intuitionist mathematics is not sufficient for interpreting the Aristotelian conception of the continuum, because for Aristotle the continuum is not a mathematical but a physical phenomenon.[6] Infinity and continuity are essentially phenomena of motion; that is why the book that begins with the definition of κίνησις (Physics III) is the systematic place for their discussion. The objection natural to contemporary thought that both concepts are, after all, 'already' embedded in pure mathematics—the continuum, in particular, in geometry—misunderstands the Aristotelian way of thinking. For according to Aristotle, the mathematical objects are ἀκίνητα only because they have no existence of their own (are not χωριστά), but can be apprehended solely by means of abstraction from actual—namely, moving—things. Infinity and the continuum, in their essence, refer, however, to κίνησις and are in that sense not really mathematical concepts; nor, as far as I know, do they ever appear as such in Aristotle. For the meaning of the term 'infinite' is, after all, merely the possibility of continuing to count, to divide, to extend the length of a line. And the continuous—for example, a straight line—is merely 'infinite in potentiality'; the line does not 'consist' of infinitely many parts but, having been divided, always lends itself to further division. All these possibilities are not 'logical' but 'real' possibilities; anyone who actually counts, divides, or extends lines is performing an actual movement. δύναμις and κίνησις belong together. The present study will list only a few preliminary observations on the subject of this relationship.

We take as our starting point the πολυθρύλητα of the paradoxes of Zeno, restricting ourselves to the first and the third.[7] The first paradox states that no distance can be traversed in finite time. For before reaching the end point, one must reach the halfway point; before that, one must reach the halfway point of the first half of the distance; and so on ad infinitum. Thus within a finite time one would have to traverse infinitely many points and infinitely many parts of the distance, which is impossible. In Physics VI.2, Aristotle answers—quite correctly, even from the point of view of modern mathematics—that a finite interval of time admits of the same infinitely many divisions as a finite distance; thus a distance and a time span are finite in the same sense and infinite in the same sense, they can therefore be mapped onto each other one-to-one; and the traversing of a distance in a time span is precisely such a mapping. The third paradox states that a flying arrow is in a particular location at every moment, that at no moment therefore does

[6] The difficulties that M. Schramm (in: *Die Bedeutung der Bewegungslehre des Aristoteles für seine beiden Lösungen der zenonischen Paradoxie* [Frankfurt: V. Klostermann, 1962]) finds in the idea of the continuum developed in Physics VIII.8 stem only to a limited extent from the manifest imperfections of the proofs offered there; they are due for the most part to Schramm's failure to consider that (a) the mathematics relevant to Aristotle is intuitionist mathematics, and (b) Aristotle is not at all concerned with mathematics here. Cf. the discussion of this chapter in Wieland (op. cit., p. 302), which, though somewhat brief, deals with the essential point.

[7] Zeno's own conception does not concern us here. Only Aristotle's conception, and our understanding or misunderstanding thereof, is the subject of our discussion.

it change its location, and thus it is altogether at rest. In Physics VI.9, Aristotle answers that motion does not occur in a now (νῦν), but over a time (χρόνος); time, however, does not consist of nows, but of times. Essential here is the definition of the continuum as that which can be divided again and again without limit into what is in turn also divisible (λέγω δὲ συνεχές τὸ διαιρετὸν εἰς αἰεὶ διαιρετά. Physics VI.2, 232 b 24–25). Time does not consist of points in time, for these are merely the limits of the 'times' (χρόνοι) or, as we would say, of the time spans; and motion occurs not in the now, but always in a time. Here, too, a contemporary mathematician, and especially a physicist, will raise no objection. In the case of differentiable motion one can, to be sure, assign a speed to a particular point in time. But speed as a differential quotient is by definition a limiting value. To define it, one must consider a sequence of time spans of diminishing duration. 'Actual', i.e., measurable, speeds are the differential quotients in the sequence; the limiting value is not part of the sequence and it is only convention that designates it as the 'speed at time t'.

Aristotle himself, we are happy to note, is not yet satisfied with his resolution of the paradoxes. In Physics VIII.8 he comes back to the first paradox, aptly remarking that in the reflection just quoted he has answered the question that had been posed, but without having solved the problem (263 a 15–18). For the problem can be formulated from a purely temporal point of view, leaving distance aside: how can one actually traverse infinitely many time spans in a finite amount of time? Anyone who does this, after all, counts, so to speak, the end points of all the spans; thus he counts 'to infinity' in a finite amount of time. If one's way of thinking has been influenced by set theory, one is usually unable even to grasp the problem (as, for instance, B. M. Schramm, loc. cit.), for it would never occur to one that 'counting' really does mean counting here. It is clear that one cannot actually count to infinity, but, nevertheless, mathematics abstracts from this. For Aristotle, however, motion is not a mathematical abstraction but a reality. The physicist trained in quantum theory finds it easier to understand what Aristotle is talking about. The physicist has learned how problematic it is to assert the existence of a physical quantity if it is certain, on the grounds of natural law, that it cannot be measured.

It is clearly impossible to 'measure' in a finite span of time infinitely many points in time—by reading them off a clock for instance. What, then, does the assertion of their existence mean? To this question, Aristotle has an exact answer: the points in time exist only potentially (δυνάμει). Any one of them can be measured, to measure all of them is impossible. What exists in actuality (ἐντελεχείᾳ) is the entire continuum (the undivided length of line; the full span of time of a real, completed process). The points on a line, the nows in a span of time during which some motion is observable, do not really exist unless they are actualized. An actual motion therefore 'counts', for example, those points that are marked by an objective feature of this motion, such as a moment of rest or a reversal in direction; of these points, however, there are only a finite number. As a student of Bohr and Heisenberg, I cannot but express my admiration for a way of thinking that is so sound from the point of view of physics.

Of course, the contemporary physicist also knows that a 'sound way of thinking' does not yet constitute a consistent theory. The manner in which Aristotle employs his way of thinking in Physics VIII to assert the primacy of circular motion has become unacceptable to us since Kepler, and, as a result, we clearly see its many logical gaps. A consistent theory of the physical continuum that embraces Aristotle's insights does not exist to this day; it would surely have to be a quantum theory. For the preparation of such a theory it may be useful to take a few further steps in analysing those problems of the Aristotelian conception which already arise on the basis of its own premises.

At issue is the interrelation, in the Aristotelian context, among time, movement, and potentiality; i.e., among χρόνος, κίνησις, and δύναμις. χρόνος and κίνησις refer to each other. Every movement occurs in a time (ἐν χρόνῳ). Time, in turn, is explicitly defined as the measure of movement according to the earlier and later (τοῦτο γάρ ἐστιν ὁ χρόνος, ἀριθμός κινήσενς κατὰ τὸ πρότερον καὶ ὕστερον. Physics IV 11, 219 b 1–2). For the paraphrasing rendition of ἀριθμός, by 'measure' the reader is referred to Wieland, *op. cit*, § 18. If one asks which of the two terms ought to emerge as the primary or more fundamental one in a systematic presentation, the answer would surely be 'movement' The definition of χρόνος uses the term κίνησις; by contrast, the soon to be quoted definition of κίνησις, requires no recourse to χρόνος. χρόνος is not 'temporality' in the full sense of the term, but only the measure assigned to it. In fact, the definition of χρόνος as a gauge for movement presupposes the 'earlier and later'. This 'earlier and later', or better 'preceding and following' (πρότερον καὶ ὕστερον), is originally, according to Physics IV 11, 219 a 15, obtained from the extent (μέγεθος) whose measure is taken by the motion; i.e., it is interpreted primarily as a spatial determination. Insofar as what we call temporality can be found in Aristotle at all, we must therefore find it in the concept of κίνησις itself.

Here, however, we are faced with a difficulty. How is the actuality of motion to be conceived of? κίνησις, after all, is the basic concept of physics. The natural thing, the φύσει ὄν, is defined as that which contains in itself a source of κίνησις (cf. Wieland, op. cit, § 15). With the turning away from the Platonic interpretation of the εἶδος the underlying reality (οὐσία)[8] was vested in these things. Thus κίνησις has become the distinguishing mark of the class of the ὄντα, i.e., of the κινητὰ—the class of primary importance to the physicist, and important also in many special investigations of metaphysics. In what way is movement?

Movement is within a span of time, but not in the now. In Physics IV, however, Aristotle begins his discourse on time with the paradox that time really is not, since part of it has passed and is therefore no longer, part of it is future and therefore not

[8] The usual translation of this term is not "underlying reality" (das eigentlich Wirkliche), but "substance".—*Translator.*

yet (τὸ μὲν γὰρ αὐτοῦ γέγονε καὶ οὐκ ἔστιν, τὸ δὲ μέλλει καὶ οὔπω ἔστιν. 217 b 33–34); moreover, the now is not time at all, it merely separates what is past from what is future (218 a 6–8). This is one of those paradoxical introductions to an investigation that are so typical of Aristotle. One can ask whether he really resolves this paradox. In the discourse on time he arrives at a definition of time as the measure of movement. Therefore, insofar as movement exists[9] time too can exist as something predicable of movement. Here, however, we find ourselves back with our question concerning in what way movement exists. In any case, it is not in the νῦν. Indeed, the νῦν itself is merely δυνάμει. A νῦν becomes actual only insofar as it is actualized by a movement. Nowhere is it stated, on the other hand, why it shouldn't be true that the past is no longer, the future not yet, that both therefore are not. If this were true, there would be nothing temporal at all: neither past nor future, since neither of them 'are here' (cf. Heidegger's convincing interpretation of ὄν as 'having presence'[10]); nor would there be spans of time, since they consist of what is past and/or what is future; nor movement, since it is only in a span of time; nor the now, because it is actual only in movement. This would mean the end of Aristotle's entire philosophy; if anyone would win out by this, it would be the 'friend of the Ideas' in The Sophist—not to mention Parmenides himself, as Aristotle interprets him.

It is worthwhile to live with this difficulty for some time, longer than the time it takes to read this Note. To prolong this preoccupation, it pays to examine a few possible solutions.

One can try to endow the present with temporal extendedness.[11] To be sure, this is incompatible with what Aristotle clearly states, but one could, after all, decide to deviate from his text. The extendedness could be an indivisible, a definite divisible, or an indefinite divisible span of time. That indivisible spans of time are of no use Aristotle himself shows exhaustively, from VI 2, 232 b 24 again and again up to VIII 8, 263 b 27–32. The strongest argument for our purposes appears in the last of the passages just listed—namely, that there is as little movement in an indivisible span of time as in a point of time; a change during such a span of time would fill it with changing contents, i.e., would divide it. A definitely divisible span of time would be delimited by points of time that would have to occur one after another and yet also be co-present. This would totally contradict the Aristotelian concept of time; for it is the very definiteness of the span of time which allows us to recognize its parts as separate and therefore incapable of being

[9] Although it seemed important elsewhere in this chapter to render the German auxiliary verb "zu sein" consistently with "to be", I have here (and immediately below) substituted "to exist".—*Translator.*

[10] "Having presence" is the standard English translation of anwesend (in Being and Time).—*Translator.*

[11] On these problems, see G. Böhme: *Über die Zeitmodi* (Göttingen: Vandenhoeck & Ruprecht, 1966).

simultaneous. In an indefinite span of time, this separation might possibly exist only δυνάμει. Aristotle did not discuss this; perhaps it is a fruitful assumption for contemporary physics. In any case, this assumption also leads to problems, ones that are analogous to those we shall presently discuss.

St. Augustine, in the 11th Book of his Confessions,[12] takes up this problem and solves it by transposing the being of past and future into the soul: memory is the presence of the past, anticipation the presence of the future. Admittedly, we will not be able to escape the paradox if we do not introduce concepts such as the presence of the past and future.[13] But their introduction by St. Augustine is un-Aristotelian as well as useless for the physicist in precisely that feature which the modern history of ideas has frequently praised as the 'discovery of the subjectivity'—namely, in the transposing of this presence into the Self. Aristotle shares with the common-sense attitude of the physicist[14] the presupposition that the movements of the things can themselves be present, past, and future; thus he requires the presence of future and past in the πράγματα and not only in the soul of the individual man. It is no contradiction that, very cautiously, Aristotle relates time as the measure of movement to the soul (ψυχὴ καὶ ψυχῆς νοῦς); he does so in case it should turn out that it is only the soul that can count. Siding with Wieland, (op. cit., p. 328, footnote 15), we can interpret this by saying that movement is outside of the soul and merely its being counted is in the individual soul; or, against Wieland, we can interpret it in terms of the world soul. In neither case do the earlier and later and movement itself exist exclusively in the consciousness of the individual man.

We have now reflected on the problem sufficiently to be able to study the Aristotelian definition of movement: ἡ τοῦ δυνάμει ὄντος ἐντελέχεια ᾗ τοιοῦτον κίνησίς ἐστιν (Physics III 1, 201 a 10–11). It is impossible to render this sentence in another language without first interpreting it. If we decide to translate ἐντελέχεια and δύναμις with 'actuality' and 'potentiality', then Aristotle is evidently saying: the actuality of potential being, qua potential being, is movement. This may seem obscure. Ross[15] tries to clear up the obscurity by para-phrasing: "Change may now be defined as the actualization of the potential as such", and in the commentary (p. 537) he explains: "ἐντελέχεια must here mean 'actualization', not 'actuality'; it is the passage from potentiality to actuality that is κίνησις". This strikes me as philologically untenable: as far as I can see,

[12] For essential instruction on this topic I am indebted to a seminar report by U. Duchrow, and to G. Picht's comments on it.

[13] Cf. Picht (1958).

[14] I cannot explain here to what extent the theory of relativity, contrary to a wide-spread opinion, has left this fact unaltered. Cf. C. F. v. Weizsäcker: *The Unity of Nature* (New York: Farrar Straus Giroux, 1980), pp. 101–137; it was translated by Francis J. Zucker from: *Die Einheit der Natur* (Munich: Hanser, 1971): II.1.

[15] See Ross (1936).

ἐντελέχεια everywhere signifies a state, and not the passage into that state; furthermore, this definition is circular,[16] since 'passage' is certainly a form of 'change'. Nor can the definition intend a coexistence of δύναμις and ἐντελέχεια during the κίνησις.[17] The τοιοῦτον would remain unexplained, and a symmetry between being potential and actual, of which there is no indication in the text, would obtrude itself into the definition. The ἐντελέχεια is clearly that of the δυνάμει ὄν or, as Wieland (op. cit., p. 298, footnote 26) correctly says: "here the modal categories are being applied to each other in a hierarchical sequence".[18] As though Aristotle had foreseen all these misunderstandings, he distinguishes two meanings of actuality (here, in a loose inter-change with ἐντελέχεια, designated by ἐνέργεια): "For the actuality of what is capable of being built (τοὺ οἰκοδομητοὺ) is either the build-ing of the house (ἡ οἰκοδόμησις) or the house itself (ἡ οἰκία). But if it is the house, then it is no longer what is capable of being built. What is built, however, is what is capable of being built; therefore the building of the house must be its actuality" (201 b 10–13). Evidently the actuality of potentially being is not the result of the movement, but precisely the movement itself.[19]

Movement, the basic concept of physics, is therefore referred back to the terminological pair δύναμις—ἐντελέχεια. This pair thus seems to be more fundamental than the concept of movement, or at least equally fundamental. What do these two terms themselves mean, and what is the meaning of their peculiarly iterated application in the definition of movement? An interpretation of Metaphysics, Book Θ, which is called for at this point, would go far beyond the scope of this Note. We must be satisfied with one particular consideration.

However one may wish to define it in detail, potentiality, δύναμις, deals with the future. The sperm is δυνάμει a human being[20]; i.e., the sperm will perhaps someday be a human being. If the sperm is a human being at all, then in the future.

---

[16] Already St. Thomas (in: Ph. M. Maggiolo (Ed.): St. *Thomae Aquinatis in octo libros physicorum Aristotelis expositio* [Turin, 1954], p. 144) offers a corresponding critique of analogous attempts in defining terms (which, however, were not intended as translations from Aristotle): "motus est exitus de potentia in actum non subito"; on which St. Thomas: "qui in definiendo errasse inveniuntur, eo quod in definitione motus posuerunt quaedam quae sunt posteriora motu: exitus enim est quaedam species motus…".

[17] Thus M. Schramm, op. cit., p. 106: "… if what has not yet been actualized were designated by δυνάμει ὄν and what has been actualized by ἐντελεχείᾳ ὄν… , then δυνάμει ὄν and ἐντελέχεια would temporally coincide during the movement, and the resulting definition would be rather apt…".

[18] Cf. also 201 b 31–33: ἥ δὲ κίνησις ἐνέργεια τίς δοκεῖ, ἀτελὴς δέ. αἴτιον δ'ὅτι ἀτελές τὸ δυνατὸν οὗ ἐστιν ἐνέργεα.

[19] The older interpreters see clearly on this point. Cf. especially the very accurate interpretation in F. Brentano: *On the Several Senses of Being in Aristotle*, translated by Rolf George (Berkeley: University of California Press, 1975).

[20] I am permitting myself this example although, according to Metaphysics Θ 1049 a 4, the sperm is 'not yet' δυναμει; the distinction involved here will be discussed in the second paragraph below.

But the future is uncertain, which explains the 'perhaps.' The δύναμις, however, is not simply the future, but that of the future which is, so to speak, already here now, and thus in a certain sense the presence of the future. An actual human being will—perhaps—be here in the future. If it is the case that he will be here, then something is already here now which is the very thing that, in the future, will be the human being, namely the sperm. The sperm is not yet the human being, though, it is only the potentiality of a human being. It is the way in which the not-yet can be now: it is the presence of the future. However, it is this not as the sperm drop it actually, ἐνεργείᾳ, now is, but only insofar as it is δυνάμει something else, namely a human being.

A short digression may be in order. A 'teleological world view' is clearly not necessary for the interpretation of the concept of δύναμις.[21] It is only necessary that there should be something like the presence of the future at all, or, to put it logically, that it be possible to infer from the present to the future. Aristotle employs teleological language, first, in areas in which the phenomena urge this language upon us today, too (in biology, for instance); and secondly, to express the overarching relations among phenomena which today are referred to as the universal laws of nature. What matters, though, is neither the language nor the world view, but the unity of time as a phenomenon (cf. Picht, op. cit.); i.e., in the problem at hand, the presence of the future.

The δύναμις may be close to, or remote from, what it is capable of becoming. Thus there is the δύναμις of a particular piece of knowledge in one who does not possess it but can acquire it, and in one who has acquired it but does not think of it (Physics VIII 4, 255 a 33–34, and De Anima 417 a 22ff.). There is also a δυνάμει ὄν that could become something definite but, until now, has not set out to become it: for example, a stone that could fall but is positioned so securely that it does not fall; or the sperm prior to conception; or the gifted idler, who refuses to study. In these instances, the future in a certain sense is not present, it is 'untouched' future. But another δυνάμει ὄν establishes the connection to the future: the stone falls, the sperm grows into an embryo, the student studies. This is the ἐνέργεια of the δυνάμει ὄν ᾗ τοιοῦτον. It is now that the δύναμις is the actual, the completed δύναμις; it does its work (ἐνέργεια), it has reached its τέλος as δύναμις (ἐντελέχεια). This is the true presence of the future. And it is this that is κίνησις.

[21] Cf. Wieland, op. cit, §16. I may perhaps remark that the profit to be derived from Wieland's book is not diminished if the reader disagrees with the sharp contrast drawn between language analysis and metaphysical profundity (e.g., pp. 139 and 179). I feel that this confrontation does not measure up to Wieland's other insights.

Thus, Aristotle defines movement in such manner that the unity of time finds its expression in it. Movement can in fact only be ἐν χρόνῳ and yet now if a future χρόνος is present in the now; the modality of this presence is the δύναμις, and the actuality of this δύναμις is movement.

In this formulation the future seems to be given priority over the past. What of the presence of the past? Now the relationship of future and past is not symmetrical; in a way that defies paraphrase, the German language expresses this asymmetry by saying: this 'wird sein', but that 'ist gewesen'.[22] The asymmetry is already expressed in the conceptual pair δύναμις-ἐνέργεια, in the concepts of τέλος and ἀγαθόν, and in ὅθεν ἡ κίνησις, and it should not be eliminated. Aristotle's doctrine of the functional and temporal primacy of ἐνέργεια over δύναμις (ἄνθρωπος γὰρ ἄνθρωπον γεννᾷ) restores equilibrium, however. Existing ἐνέργεια presupposes either eternal being, as with God, or a prior becoming. An ἐνέργεια that has become is necessary for a new δύναμις; a sperm is the sperm of a future human being, but it must be the sperm of a grownup man. The grownup man is the result of past growth, he is its presence; i.e., the presence of past δύναμις insofar as it has become actual as κίνηςις. The actuality appears in the form of perfection, of ἐντελέχεια μή τοῦ δυνάμει ὄντος ἀνθρώπου, ἀλλὰ τοῦ ἀνθρώπου ἁπλῶς. In this sense, the presence of the future presupposes the presence of the past.

In conclusion, the author must ask himself whether these are Aristotle's explicit thoughts. An interpreter who himself philosophizes is constantly in danger of doing violence to the text. The experience of frequently finding Aristotle easier to understand than the difficulties discovered by his modern interpreters is a source of encouragement, and it may also be a source of error. In any case, I wish to stress my adherence in this Note to the following principle of interpretation. I would not have turned to the study of ancient philosophy had I not been baffled by unintelligible terms in the conceptual tradition of modern physics, the humanities, and modern philosophy that I could only hope to understand by going back to their historical sources. Indeed it seems to me that the great progress in the modern era —such as the rise of exact science, the development of subjectivity, and the growth of a historical awareness—has been achieved at the cost of a certain narrowness in the questions asked and the conceptualization employed; to put it more pointedly, at the cost of a progressive 'Seinsvergessenheit'.[23] In probing the foundations of any domain of contemporary problems—e.g., of physics—one discovers the same structures that the Greek philosophers discovered long ago, if from a different angle. This is why work on contemporary problems might be helpful in understanding Plato and Aristotle, and vice versa.

[22] "This will be, but that has been"; unlike in English, the past tense in German is constructed with the present tense of the auxiliary verb 'to be'.—*Translator*.

[23] A Heideggerian term (in Being and Time) meaning "forgottenness of being".—*Translator*.

References

Picht, G., 1958: *Die Erfahrung der Geschichte* (Frankfurt: Klostermann), Section VI.
Ross, W. D. (Ed.), 1936: *Aristotle's Physics* (Oxford: Clarendon Press): 359.
Wieland, W., 1962: *Die aristotelische Physik* (Göttingen: Vandenhoeck und Ruprecht), Sect. 17.

Carl Friedrich von Weizsäcker at the age of 82, photograph by Lili Bansa in 1994. © Lili Bansa who granted permission to use this photo

Chapter 10
The Rationality of Emotions

10.1 The Concept of Interest

What leads us to inquire about the rationality of emotions?[1] Perhaps, at first, scepticism about the usual view of their irrationality. This view likes to pair the concepts of rational-irrational.

What do we mean by that? If one says about people that they behaved rationally, one means, perhaps, that their behaviour complied with two conditions:

1. One can give fairly accurate information about the causes of one's behaviour, and
2. in view of one's own interest and of the interpersonal values, one can feel somewhat compatible with one's own behaviour. In Habermas' diction, it could be defended in a practical discourse.

Condition (2) contains two criteria: personal interest, and interpersonal values. Common and perhaps even dominant, however, is a rational pattern of behaviour in a limited sense. People acting with complete consciousness realize their own interests even in opposition to interpersonal values of which they are aware and which they do not oppose. We could call this mode of behaviour 'rational interest'. It is not completely rational insofar as it contains the pending conflict between our interests and established values. One can even maintain that a function of the

[1] This was originally a private note (A.) and an internal paper "Conditions of Life" (B.-G.) of the 'Max-Planck-Institut zur Erforschung der Lebensbedingungen der wissenschaftlich-technischen Welt' ('Max-Planck-Institute for Research on the Conditions of Life in the Scientific-Technical World'). This text was first published in: *The Ambivalence of Progress* (New York: Paragon House, 1988), pp. 142–160; translation of: *Der Garten des Menschlichen* (Munich: Hanser, 1977): II,4.

M. Drieschner (ed.), *Carl Friedrich von Weizsäcker: Major Texts in Philosophy*, SpringerBriefs on Pioneers in Science and Practice 23,
DOI: 10.1007/978-3-319-03671-7_10,

public acknowledgment of interpersonal values is at least to maintain this conflict and not always to give into private interests. The conflict is also a conflict of the personal interest with itself because the interpersonal values reflect the interests of the community whose well-being is a condition for the well-being of the individual. If the individual thinks rationally, one has to modify one's understanding of one's own interests to adjust to those of the community interest. Here we see the bottomless nature of a purely empirical concept of interest based on the opinion of the individual. Its contents depend upon a judgment that might be modified after further reflection. The question arises as to what is the real interest of both the individual and the community. Even if the individual convinces himself, as is usually the case, that pursuit of one's own interests serves as well the common interest, he argues with a concept of the real interest of the community.

If one observes the real effects of rational interest behaviour, including the destruction of the environment, and wars caused by powerful interest groups, then it does not seem an exaggeration to speak of the irrationality of the rational. Perhaps one should more precisely state that the practical rationality needs practical reason, that conceptual thought of interest needs a perception of whole that only makes recognizable the real interests of the individual and the group. In the light of this analysis, condition (1), above, reveals its meaning. What criterion do we have for determining if we understand the reasons for our own behaviour?

One direction in conflict research finds the reason for conflicts, or at least their insolvability, in the irrational motives of our actions.[2] These motives, such as aggression and fear, are certainly important. But one can understand their importance only if one recognizes them as deviations from some sensible behaviour. One could term the next goal of understanding in this context, "the insight into the rationality of the irrational".[3] Sensible 'irrational' behaviour can be called a perception of reality without concept. Animal behaviour is certainly like this. But in human culture, conceptless perception of reality plays an even greater role than interest rationality is aware. The rationally unavoidable perception of self-interest is a good example. But also the rational rationalization of individual interest as the real interest is based on a highly differentiated conceptless, or not fully conceptualized, perception.

At this point, a brief look at a contemporarily neglected range of experience might be useful. The most explosive interests may be identifiable by the fact that we work hard to control them. The most obvious and most symbolic renunciation

---

[2] Cf. *The Politics of Peril* (New York: Seabury Press, 1978); Chap. 7: "On the Theory of Power", pp. 142–155; reprinted as Chap. 9 in: Ulrich Bartosch (ed.): *Carl Friedrich von Weizsðcker: Major Texts in Politics and Peace Reasearch* (Cham et al.: Springer-Verlag, 2014), translated by Michael Shaw from: *Wege in der Gefahr* (Munich: Hanser, 1984), Chap. 7.

[3] Cf. *The Ambivalence of Progress* (New York: Paragon House, 1988); translation of: *Der Garten des Menschlichen* (Munich: Hanser, 1977); Chap. I7 ('Beauty'): 96–102.

of interest is probably that of the mendicant friars. The three monastic vows are: poverty, obedience, and chastity, or renunciation of property, of domination, and of sexuality. Of course these are also the most explosive interests.

10.2 Place and Action (Possession, Domination, Power)

The human being is a living being. Every living being is a body. Everybody is in a place. A person needs at least a place.

Life is a process. To this process belongs at least metabolism, consequently nourishment. The higher animals get their nourishment by moving within a territory. A human also has to move. We call his movement action if it is meaningfully coordinated. One's place is not only where one stays, it is also the necessary territory for one's activities.

Higher animals usually live in societies. The absolute minimum requirement is temporary socialization of the sexes necessary for mating. Society is fundamental for humans because we owe to society the wealth of behaviour that we exhibit in language and customs, without which we could not live. For social life this treasure of behavioural patterns in respect to other members of the species is fundamental. This first gives the individual a 'place' in the society that is also territory for acting. Because of this analogy, I entitle this section with the abstractions of place and action.

The elementary possession, which already exists in animal societies, is itself a place; the ecologist would call it territory. Behavioural scientists have shown us that a fish bravely defends its territory, and they inform us that the male nightingale sings so that others know which is its territory and refrain from entry. It appears fundamental that evaluation of the human grasp at possessions must include a very long animal prehistory of territorial possession.

Territory possession is the animal habitat. Territoriality, however, is not universal with animals; but it is a usually acquired behaviour pattern. The territory commonly does not belong to the individual but to the social group and sometimes, in a distinct sense, to its leader. The readiness for territorial behaviour is always present in humans. The transition from the cultural forms of gatherer, hunter, nomad, farmer, urban dweller only give different forms to the disposition of land, as does the varying size of the group that participates in the possession. It is rather a simplification when some social theories assume that at one time the most powerful divided what had been common to all into private property. The historical process vacillates between both tendencies. Especially at the highest levels of culture, and accordingly individualization, there is recognition of property held in common as an achievement of a society that imposes on its members a high standard of asceticism. I believe that humankind's ability to renounce landed property is possible only if one has another home, an unquestioned place in the society, or a spiritual existence.

The second form of property, that of things, has only minor precedents in animal life. In humans it is related to the capacity for methodical action. The first tangible possession of things is a means to an end: food supplies, tools, weapons. What is fundamental here is not habitation, thus existence, but ability. To be able means to be able whenever one wants, accordingly power. With power, probably the most important ambivalent attribute enters the picture. I pass over the 'territorial' relationship that extends also to one's own tools, to one's own furniture, as well as the dominant power function of real property as it exists in advanced power societies. I rather seek to approach the phenomenon of power in a direct and abstract way. To do so, it is first necessary to differentiate between power and domination.

Social hierarchy among members of the same species and domination over members of the same species is as common among higher animals as is possession. One can say that these relationships of domination and subordination are fundamental to animal societies with generally individualized relationships between their members. As far as humankind has been determined by its animal heritage, this is valid for humans, too. It is again an optical illusion of certain social theories to maintain that domination and hierarchy are the result of power that is contrary to nature. Through nature and custom, the human being has been conditioned to secure his or her social standing exactly by knowing who is above and who is below. It is an extraordinary historical effort and achievement, where it succeeds, when people indeed recognize each other as equal. Later we must ask how this achievement is possible at all.

Animal societies, and also human societies to the extent that they have similar dominant traits, I will call simple societies or societies of actuality. Let us imagine someone who knew only such societies and who was then introduced to the realm of power. He would find himself confronted with completely strange phenomena not deducible from his previously existing knowledge, and with a new level of behaviour. Despite the continual quarrel over hierarchy and possession, they existed within the narrow limits of the actuality of being and having. Now, though, one is confronted by the essentially unlimited realm of ability or power. I use the term power here as discovered potentiality. Potentiality is never completely and definitively realized. Complete realization means the end of the potential. Power is principally understood as the ability to do what one wants to do and as not having to do all that of which one is capable. That such a realm of possibility exists at all is a discovery, and it belongs in this respect under the heading of insight, but it can be observed here already as an isolated factor.

Power through tools, hunting weapons, and habitation is, at first, power over nature. I am not referring now to the well-known exterior process of humankind's seizure of power on earth, but about its effect on human patterns of behaviour. Humankind must have been deeply transformed and affected at the beginning of history by the discovery of power. The fascination and horror of this transformation are still reflected in power dreams of magic and in the religious reconciliation of the powers of nature, a striving for the reestablishment of equilibrium. I quote three testimonials from classical high civilizations: the quiet command of God in the biblical Priestly Source to "have dominion over all the earth", the deeply troubled chorus in Sophocles' Antigone, »πολλὰ τὰ δεινὰ«, and Chuang Tzu's anecdote of the wise peasant who did not want to use the bucket wheel at the well because, "Whoever uses a machine gets the soul of a machine."

Power, though, obtains its actual explosive effect as power over others, through weapons, through superior knowledge, through the functional structure of advanced civilization. The specific phenomenon of domination that is the subject of the contemporary debate arose from the traditional social hierarchy because of power, from the natural fights between the groups and within the group in the art of war and the art of politics. At this point we are close to the source of the ambivalence of progress for the first time. Also, the power over people, once it is discovered as a possibility, is subject to no immanent limitations. If much is possible, then more is possible. Only one who wants to make the second million will make the first. Here possession is no longer territorial security nor fulfilment of needs nor a means toward luxury. But it is the unlimited scope of activity, it is power. Therefore the requirement that profit be measured in accordance with needs completely fails to understand the principle behind profit maximization. The same applies to concentration of power in the hands of the state. It is from this structure of power that I argue that I find a Third World War probable and that I think world government is the most conservative solution. At least this solution does not attempt the colossal task of breaking the power of power.

Jakob Burckhardt says that power in itself is evil. I always resisted this statement, which gave frightened aesthetes the justification for running away from the responsibility of politics. But I was just as much fundamentally dissatisfied with the claim that power in the right hands was good, even if this often is practically the most tolerable compromise. The self-appointed 'right hands' of the conservatives as well as the revolutionaries characterize the problem. Occasionally I have rephrased the thought "Power, as such, is evil," namely, power not integrated into the framework of human values. But this is merely another embellishment because its not being integrated only signifies the discovery of the unlimited potential. I would be inclined to say today that power, even if combined with good will and responsibility, is ultimately tragic. It has to render its opponent powerless, harmless, or it will not last, and a result of power is that it produces opponents. But the victors, the successful bearers of power, discover their own deep powerlessness, their own incapacity to change by means of power the basic structure of power. Tragedy is accompanied by delusion: Whoever still fights for power will hardly notice this powerlessness.

Tragedy is not, however, the last word. I have only taken an isolated look at power. We must now come out of isolation.

10.3 Love

Let us imagine someone who only knew the realm of possession, domination, and power, and then meets love. One would see oneself again confronted by a completely strange phenomenon, not deducible from his existing knowledge.

Possession, domination, and power are domains of the ego. Even the relations between groups in these domains are known appropriately as group egoism. Interest rationality serves interests that the ego knows. The tragedy of power is the impossibility to satisfy the ego. We experience in love a fulfilment of the ego that is completely incomprehensible from interest rationality, where the ego is overcome, a transformation of all values. A science that thinks instrumentally can easily describe possession and domination because the very concepts of science are in the service of such finite rationality. The limitlessness of power finds its partner in the limitlessness of knowledge. It is almost impossible to determine love scientifically because it transcends the very determinant. In spite of this we shall attempt the task.

That living creatures must have descendants is biologically not as self-evident as it seems. It is also related to progress. In an eternal world, in a cyclical time, the continuous existence of the same individuals would be imaginable. This is the dream of personal immortality. Life that newly emerges needs descendants to populate space. Accordingly, evolution needs new individuals so that there can be new species. Species with a short life span for individuals and with many descendants will develop faster, and therefore survive in the struggle for existence. Sexual procreation, when understood from selection theory, appears to offer the advantage of the accumulation of recessive characteristics. Accordingly, this is how this highly perplexing form of procreation a priori may have developed and become successful.

From a biological perspective, securing sexual procreation therefore safeguards an interest that goes far beyond the needs of the individual or the individual group. Our concept of interest rationality, however, is in accordance with the needs of individuals and groups. The sexual instinct must forcefully break this world of understandable needs. It cannot, and consequently must not, argue. This might explain some of its enchanting power. Among animals it creates the richest rituals. It luxuriates in a manner that makes its productions akin to those of the freely playing mind. Eroticism moves away from the sole purpose of procreation in two directions: toward a personal bond and toward purposeless beauty. The strongest non-rational force in humankind becomes the bearer of a development that is only made possible by overcoming mere interest rationality.

Against this background can be seen the ambivalent struggle of all cultures with sexuality. Sexuality, even if completely different from purposeful rational power, is a domain of the possible, too. The plasticity of human behaviour permits many different ritualizations of eroticism. But for these ritualizations to be successful, there must be sufficient pressure for their realization. Therefore, the expression of sexual instinct that runs counter to rituals has to be repressed. To name a classical example, one must avoid the polygamous tendency for the sake of the development of one personal bond and the care of children. What is being repressed, however, keeps the radiant power of being superior to all the narrower interests: the prospective release of the ego from its own boundaries. The error is repeated again and again that this force could even then be preserved when it is given the status of a positive ritual. But when thus acknowledged, it enters the sphere of interests. The test of its value is then only its constructive force in that sphere. Tabus, being blind and not knowing their own purpose, have exactly that in common with instinct, and to play one against the other is, at times, considered to be enlightened.

Given that, it comes as no surprise that science discovered the mechanism of repression exactly in the example of sexuality. Repression here appears as a condition for the precision of a ritualization, and sexuality is the instinct that most encourages ritualization. But to understand better the meaning of repression—obviously a process fundamental for ambivalence—we have to ask what is the conscious and what is the unconscious. However, we cannot do this until we have examined insight.

Bur first we must remind ourselves that there exists still another completely different relationship between individuals that is also sometimes called love, the nonsexual personal bond. Because it interweaves with all communal life, we have come across it many times already without, however, it becoming a theme itself. Its quality, though, has to appear as not deducible from hierarchy, power, and sexual love. Every direct reductionism from either sexual theory or the theory of purposeful rational behaviour fails to explain it. The most interesting book about its animal prehistory is, in my opinion, Konrad Lorenz's *On Aggression,*[4] especially the chapter "Das Band" (The Bond), which to me is more interesting than all of the other ideas expressed in his book, which have become more popular. Lorenz genetically traces the personal bond back to aggression and its assimilation through ritualization, through which he sees the possibility of individual relationships within a group that goes far beyond the first achievement of ritualized aggression in the form of the social hierarchy. There, too, the ritual does not realize self-evident values but creates or makes possible values that are not self-evident. If Lorenz is right, then he makes comprehensible the structural relationship of various kinds of personal bonds, even if these cover a very wide area ranging from indissoluble friendship to partnership and companionship to lifelong animosity. This appears to me as one of the important contributions to the understanding of ambivalence. It is especially a contribution to the comprehension

[4] Konrad Lorenz: *On aggression* (London: Methuen & Co., 1966); many other publishers. Translated by Marjorie Latzke from *German: Das sogenannte B—se* [The So-Called Evil] (Wien: Borotha-Schoeler, 1963).

of institutions that are neither interest-rational nor sexually reducible, such as marriage.

Obviously, the personal bonding of humans, just like all other relationships and probably more so, is in its gestalt comprehensible only through insight.

10.4 Insight

Let us imagine someone who only knows possession and hierarchy, power, love, and individual bonding. Could this person understand insight? In a certain sense, however, we cannot imagine this person at all because to know all these things must mean having insight into them. We cannot even imagine someone completely without insight. But if one, as is customary, knows interests, instincts, and order by having adapted to them, does one not already have an insight into one's insight? And could such a person comprehend what insight is?

To make insight understood, we would need a philosophical psychology that, as far as I know, does not currently exist. The empiricists, who dominate science, do not usually see the problem, and the philosophers, who should be educated to see the problem, cannot recognize it in the thought forms of empiricism.

Insight is not self-training through trial and error, but rather the ability to anticipate the success of each act. It is, after all, 'insight' into the reason of such successes. It might be a very partial insight or a comprehensive awareness, it always rests on what I can only call 'truth'. Truth, of course, does not mean infallibility. Terms like fallibility or error, also lie, only make sense where it is already understood what is meant by truth, even if not reflected upon. Insight means being determined through some recognized truth, through a truth that has proven itself to the individual. This begins with instruments. An instrument consciously constructed and employed is not possible without insight. It continues with sociability. Social customs might be passed on and observed without insight, although I doubt that they emerge without some kind of insight. There is no personal relationship without insight. How far one can proceed with insight cannot be stated beforehand because every special insight permits further inquiry into the cause of its possibility, which, when answered, offers insight on another level.

Insight seems to cut across all compulsions of instinct, habit, 'unreasonable' interests, and custom. This diagonal path is connected with its disclosure of the three related regions of facts, possibility, and freedom. I can free myself from a compulsion only when I can confront what compels me as a fact. In this sense we can say: "What wonderful things facts are." The fact, as a fact, as something that happened, is unalterable. To not see the facts, 'wishful thinking', is self-inflicted bondage. To this extent, innate behaviour, individual habit, and social custom are not free no matter how well they may be adapted to the circumstances. Insight into facts liberates insofar as it opens possibilities. The recognized and acknowledged fact is therefore no longer a compulsion; its state can perhaps be altered. Insight is insight into possibilities. A fact is understood as exactly this fact only by being thought of as a possible fact and thereby being acknowledged as something that happened. Thereby, however, other possible facts are set up as possibilities. These possibilities are not fantasies but 'de facto possible' insofar as they are based on recognized facts. To have factual possibilities, however, means freedom.

These observations are philosophically abstract. One can speak concretely about insight only by speaking of the concrete facts and possibilities into which there is insight. Thus our anthropology will lead to concrete observation of history; it cannot do otherwise. First, though, we have to pursue the abstract structures a little further.

Insight not only opens freedom, it also needs freedom. This is valid psychologically. If I look at the state of consciousness of a person who experiences this state as insight, and if I understand its psychological cause, then I cancel its characteristic as insight. This is the figure of criticism of ideologies. If my 'insight' is determined through my economic interest (or my libido, my neurosis, my tradition, or my contrariness), then this causal derivation is unmasked as not being insight. This unmasking is affected even if the 'insight' concerned corresponds with the facts, consequently fulfilling the classical criterion of truth. Because this correspondence is then 'accidental'; and only the critic of ideologies who deliberately examines it has (at least he thinks so himself) true insight. I will not pursue here the resulting philosophical problem of truth and freedom, but rather turn now to the ethical-social question.

If human life is not possible without insight, then common human life is not possible without common insight. This is my meaning when I say that a peace is the embodiment of a truth, truth the soul of peace. The understanding here of peace is the possibility of living together. I speak of 'a' peace because there have, historically, been many forms of common life. Of course, there is an ethical claim in this use of the word 'peace'. The fact of common life can also embrace fighting, wife-beating, and, between two superpowers, the waging of war against each other. Even in such coexistence one usually can still find the rudiments of a peace, e.g., the continuing common life of husband and wife, the framework of martial law, and the continuation of politics embracing war objectives that envision future peace. But peace always signifies, first of all, the value for which we strive and whose absence signifies the terrible misery of modern times. The peace for which we strive is not mere coexistence but what I call, in a terse sense of the word, 'possibility', the possibility of living together. It is the possibility of life common to humans that opens a common freedom. This freedom requires a common insight, consequently a common truth.

One can say now that truth is common by nature. Two people who recognize the same thing are united in this insight beyond all arbitrariness: They cannot shake off this mutuality even if they wanted to. They have common facts and the resulting common possibilities. Therefore, I have stated earlier that truth as such is intolerant. Recognized truth excludes the possibility of sincere acceptance of the opposing untruth. In another sense, the orientation in reference to truth is, by nature, tolerant. The mutuality of a certain truth that I seek with my neighbour requires that he understands this truth, that it be truth for him. The mere repetition of a statement that expresses my truth is, however, not truth to him even if he thinks he believes what was said. If I want to live with my neighbour in a mutuality obtained through truth, then I have to permit him the freedom of approving this truth. That is, I have to grant him peace. In this regard, peace is not only a consequence of comprehended truth, but at the same time a precondition for its factual realization. In this double meaning, I call peace the embodiment of truth.

Obviously, in practical experience, many problems emerge concerning education, clarification, and substitution. This is obvious and need not be further discussed. I only point out a kind of tension that is another moulding of the tension existing between tolerance and intolerance. All mere factual authorities become invalid in the name of truth and in the light of recognized truth. What makes the achievement of equality among humans possible is always a common truth. On the other side, one authority is immanent in peace based upon truth, namely, the authority of whoever possesses insight. Truth equalizes its perceivers. Social equality in the name of truth can also go so far that it considers everyone equal and as virtual perceivers of this insight, especially if nobody claims to possess a monopoly of the truth himself as an institutional basis for power. But this equalization is false if it denies the difference between actual insight and the lack of insight. Such a fiction will be unmasked in the course of time because of the instability of the peace that is based upon it. Such a fictitious equalization makes both partners equally unhappy. For the one lacking insight, the most important insight accessible to him is the insight that he does not have the insight in question; any equalization that obstructs his access to the knowledge of his ignorance does not take him seriously as a person virtually having insight, therefore it despises him while claiming to respect him. On the other hand, the person who knows is forced, because of this fictitious equalization, to deny, to the person who does not know, this indication of respect and consequently any help the other could be given. I have never found people as free as where the authority of insight was acknowledged as self-evident.

By using the indefinite article in ‘a truth’, ‘a peace’, I have begun describing the factual plurality of truths and of forms of peace. Traditional societies live under a religious truth and under the peace it makes possible. The contemporary world lives with the truth of science and with the peace of the technocracy made possible by it. These two examples will demonstrate to contemporary young intellectuals the questionability of the achieved peace and the questionability of the truth upon which it is based. Does this alone not refute the entire anthropological position on truth? If there is one truth, who possesses it? If there are many truths that are obviously contradictory in their realization, what sense does it make to call them truths?

It was against this background that Nietzsche said, "Truth is that kind of error without which a certain species would not be able to live." If 'to be able to live' means the possibility of common life, or 'a peace', using my terminology, then truth would be in accordance with Nietzsche's definition the error that enables peace. Let me anticipate at this point the philosophical question about truth with an abstract and individual psychological reflection.

The abstract reflection: In "Models of Health and Illness, Good and Evil, Truth and Falseness" (cf. Chap. 6 in this volume) I tried to see illness as a 'false health' and error as a 'false truth'. The language of cybernetics ("adjustment to a non-optimal set point") can be used to make the point but it is not essential for establishing the basic concept of 'false truth'. If one understands as truth that something appears as it is, then what appears in error is not nothing: Something does appear, but not the way it is. At least something appears, and perhaps we can even live with it for a while or in an ecological niche. Thus the error that enables peace is an incomplete truth. Only when it claims to be obviously true does it become false. But nothing is known to us in its complete truth. Our articulated insight is always 'a kind of error'. Therefore, 'truths' must oppose each other. It is precisely the legitimate claim made by errors to be truths that compels them to opposition. This opposition moves the historical process. The historical process can appear as a chain of successive truths. This is what dialectics formalizes. The historical process can be represented in our contemporary perspective only as an opening toward the unknown. We cannot explicitly think of world peace as the embodiment of the final truth; where this hope takes the shape of concrete thoughts, it evidently becomes unreal and tyrannical. We must seek to comprehend truth as the sequence of truths, as the truth of the process itself.

Let me add that the unlimited succession of truths is related to the limitlessness of power. Power is always based on an insight. Whoever can achieve peace is the most powerful. But the tendency to use peace as a vehicle for power is a non-truth, a perversion of peace, and ultimately a cause of conflict from the perspective of the possibility of being able to live in common. We must expect that we cannot easily get rid of the problem of power through progress.

The reflection of an individual psychological parallel: Even in a person's intimacy with himself there is peace and conflict. Here, too, peace is the embodiment of a truth. Here one can see very well the limited value of limited peace. The human being learns, matures. Certain insights are more natural at certain periods of life. Goethe said, "If you are someone, do not stand still; you must move on from one light to the next." This transition from one insight to the next rarely happens when one is at peace with oneself. Crises are the usual manifestations of conflicting truths within the individual. Inner peace can result as much from the happiness of having obtained insight as from the suppression of troubling thoughts. One is inclined to call the one a true peace and the other a false peace. Are there criteria for this evaluation?

10.5 Values

Can there actually be insight into values? Or is there only insight into facts that we then value 'irrationally'? Is agreement about values only mutual irrationality? If not, what is the rationality of values?

This question has been dealt with de facto from the beginning of this book. At the outset, I proceeded with a conscious naiveté from values that I assumed were held in common by all thoughtful people in our era. Insight into ambivalence undermines the naiveté of our values. To gain insight into the cause of ambivalence, we proceeded to anthropology. The question concerning insight into values is an abstract philosophical question. I pose it here intentionally, though, in an empirical anthropological context. The first attempt to answer it must not be burdened with philosophical considerations. How do we react—provided we have not entirely lost a certain naiveté in our own values—to anthropological material?

I have frequently emphasized that it is very easy to be certain (and consequently in agreement) about values, if their absence is experienced as an obvious lack of values, as obvious suffering. We can now say that in this regard we are not different from animals. Animals do not need to ask if self-preservation and preservation of the species are values. Hunger, fear, sexual love, mother love, speaking anthropologically, compel animals to do what these values demand. If animals were not so constituted, they would no longer exist. Humankind can question if all this is actually a value or even the ultimate value, but even for us this question is usually not raised in elementary situations, or it is met with silence if it is raised. It does not, however, become silent with necessity, because a human being can question the meaning of his survival. This scepticism or despair is even a natural symptom of certain crises of maturity. We also understand that human life should not exhaust itself in providing for basic needs. (The same goes for animal life, if we examine it closer.) We hope to secure our lives in order to make possible a good life. But what does 'good' mean? This is obviously the central question in the Research on the Conditions of Life[5] (ζῆν καὶ εὖ ζῆν).

The American Declaration of Independence lists among the rights of humans the "pursuit of happiness". This phrase seems to me to be almost a summary of all the ambivalence discussed here. It can be interpreted as simple truth or as the source of human misery. The commitment to such an ambivalent phrase clearly characterizes the self-contradiction of America and our American world epoch. Is happiness a value or even the criterion for other values?

Pleasure and pain are, from the biological perspective, indicators or signals of the advantageous and disadvantageous. In observing and judging animal behaviour (which animals cannot do) according to the most simple criterion of survival, we see the resulting clear hierarchy of Darwinian values. There are, until today, individual members of species that were able to develop quickly enough. From a

[5] This alludes to the name of the Max-Planck-Institute where this text originated, cf. footnote 2 —*editor* [MD].

biological perspective, progress is a higher (that is, more successful) value than the mere preservation of the individual, preservation of the individual is a higher value than pleasurability and painlessness. The lower value only expresses a condition that usually serves the next respectively higher value, but which is sacrificed (or fails in the struggle to survive) if it does not serve. Pleasure and pain, though, are a kind of value that is different from preservation and progress. On the one side, pleasure and pain can be understood as purely functional, and if something functions by itself, it does not need other value indicators. On the other side, they serve all values, not just the lowest; sexual pleasure, for example, does not serve the preservation of the individual which experiences it.

Now, using the value of mere survivability, we judge other organic life insufficiently (did dinosaurs not have a fulfilled life merely because other species won the race of evolution?). And for humankind, pleasure and pain enter into a completely new context.

There has been, in the human disposition, a disintegration of instinctive constraint that considerably devaluates the indicated function of pleasure. Civilization has largely separated elementary instincts from their original biological intent. The struggle against this separation is itself ancient. What our tradition calls animal lusts (gluttony, drunkenness, fornication, even murder should be listed here) are in reality specifically human inventions. In all of them, a pleasure that no longer serves as an indicator is structurally linked to unlimited capacity, to power. The same is true of the relief of pain through drugs. But it is especially true about the higher and more abstract forms of pleasure that domination and property provide. Pleasure is here a principle that is inherently without insight. It functions for the animal as an indicator instead of insight. Humankind, however, needs to control pleasure through insight. The pursuit of happiness, as a principle of pleasure if separated from insight, can have ambivalent, and ultimately destructive, effects.

There is a contemporary point of view that holds that the destructive effect of pleasure is only a retaliation against social repression and that if pleasure is not repressed and free to expand, it will find its limiting gestalt by itself. To me this viewpoint mixes deep insights with naive errors. It is itself an example of ambivalence. It is self-justification for experimental experiences that are probably necessary today. Its error for me is a naive optimism that is neither historically nor scientifically justified. But the positive insights are far more important for my present argument.

Again and again, in organic development and especially in human history, we see that old material can be used for completely new purposes (such as the air bladder becoming the basis of the lungs, and aggression the basis of communication). Accordingly, the intellectually, and often esthetically, stylized satisfaction of elementary instincts becomes the embodiment of culture: eating a meal together as culinary art, drinking as symposium, living in the city and the country, a society structured according to social roles, love as play, as emotion, as partnership. In many of these manifestations, one might just as well talk about insightful happiness rather than basic pleasure. One of the highest kinds of happiness is to experience progress, to see one's own productivity. Is happiness in this sense not the true human value? (“All pleasure desires eternity”, according to Nietzsche.) And will this happiness not find, through inherent insight, its own limits where that is needed?

This concept, human happiness as a dominant value, includes, by the way, the unity of value as well as the plurality of values. This plurality is part of the structure. Humankind is happier when not always doing the same thing, people are happier together if they are not all equal (‘three cheers for the little difference’). Equality is equality in freedom, freedom is the freedom to be different. But structure is based on the commonness of this plurality of values, the solidarity in acknowledged differences. And this communal happiness is a value, the dominant value, in a certain sense the value.

This dominant concept of value can be real if the conditions for the possibility of its realization can be fulfilled. Insight and creativity are among them. None of the above-mentioned cultural stylizations of basic needs are the logical result of their biological purpose. Instinct and elementary pleasure are everywhere nothing but material. What are the conditions of insight and creativity that are necessary for their formation?

We can understand happiness and suffering, to use more abstract concepts than pleasure and pain, as indicators of the success or failure of cultural processes. Perhaps suffering is the more reliable indicator. It is also the indispensable teacher and motivator. The experience of happiness has the inherent natural tendency to become shallow. If one is of the opinion that progress is a dominant value—and the unlimited number of possible insights as well as the mere concept of creativity indicates that it is—then this is be to expected, because the indicator happiness invites lingering. The struggle between competing truths is accompanied by suffering. The positivity of ambivalence consists in the insights that result from the suffering that it causes. The danger consists in accepting partial happiness without any insight. Happiness experienced by the individual cannot be the value at this stage either. The belief that suffering is only the consequence of society’s defects is naive (or a projection). How would we be capable of feeling physical and psychic pain if we did not need this indicator? Still, we will encounter the question of a painless society again.

One criticism of the contemporary world is expressed in the concept of alienation. If this means alienation from a life of fulfilment through happiness and suffering that has been replaced by one of meaningless work and functionless pleasure, the criticism is valid. But alienation from happiness would be as ambiguous a concept as that of happiness itself.

Our observations have led us back to the concept of insight as supplemented by the concept of creativity. But these concepts have remained formal—except for some brushstrokes for a picture of cultural stylizations, from culinary art to partnership. What is the truth that guides insight and expresses itself in works of creativity?

10.6 The Enlightenment

The Enlightenment is, according to Kant, humanity's taking leave of its self-imposed minority. Kant explains minority as the inability to use one's intelligence without the guidance of another. This minority state is, according to Kant, self-imposed if the cause is not lack of intelligence but is lack of resolution and courage to use intelligence without the guidance of another. Sapere aude! (Have the courage to use your own intelligence!) is thus the motto of the Enlightenment.

Kant's essay "Beantwortung der Frage: Was ist Aufklärung?" (Answer to the Question: What is the Enlightenment?), first published in 1784, was written on the special problem of freedom of expression against restrictions of the church and the absolute state. Kant's definition, however, reaches far beyond that, and I want to make it my own in an explicit interpretation that leads further.

Kant uses the concept of intelligence that sounds familiar but is always understood, of course, in light of his elaborate concept of reason. I use here the much less elaborate concept of insight as dealt with in this text, and I understand intelligence as the ability to gain insight. The state of minority then is an inability of gaining insight oneself. It is self-imposed if it is the result of a lack of courage.

But what does this have to do with the historical process of the Enlightenment? The intention here is that an entire society should emerge from this state of minority. Therefore, the parallel of the maturity of the individual becomes significant. Every person is originally minor and not yet capable of using his or her own intelligence. One must mature to reach certain insights and then summon the courage gained, shake off authority and thereby come of age. But if we make the concept of minority mean that, then it appears to us that the concept of one's own guilt slips away. Am I guilty of being born a helpless child? And, from another perspective, am I guilty for what my parents and teachers taught me? Am I guilty of the society in which I have grown up? Did not Enlightenment understand itself more as emergence from a minority not from its own fault, as liberation from a foreign bondage?

Kant's concept goes much deeper than this thesis of rebellion. He touches on a subject that is also dealt with in the attempt to heal neurosis through psychoanalysis, or rather in any self-discovery of a responsible personality. If I recognize a flaw I was caught up in, there can be no healing, no escape from the causes as long as I look outside of myself for the causes, that is, in what has been done to me, in my parents, in society, in an inner compulsion, in my unconscious, or in my instinctual structure. All of these observations might, factually, be entirely correct. But the ability to overcome my flaw and not immediately regress to the same dependency all over again is that I do recognize it as my own flaw. Mea culpa, mea maxima culpa is not an expression of false remorse, but the clear recognition that occurs exactly when I become mature and capable of responsibility. Rebellion that is directed against the real mistakes of others contains the possibility of enlightenment, but it is not yet enlightenment, because recognition of the mistakes of others permits me still to conceal my own mistakes by projecting them onto others.

The use of my own intelligence is the search for truth. Thus the endeavour of this book may be seen under the title of the Enlightenment as understood by Kant. Which truth is sought? Naturally that truth that is relevant to the previously asked questions. Which is that?

Kant's thesis exists in a political context in its broadest meaning as does the entire concept of the Enlightenment. His paper is the answer to the mistrustful and sceptical questions of a clergyman. It is an aggressive and direct Defense of the concept of the Enlightenment. Beginning with the eighteenth century, the term Enlightenment designates an intellectual movement whose tendencies are exemplified in the use of the words progress and emancipation. Enlightenment understands itself as the triumph over absolutism, and it accompanies the movements of liberalism and socialism. It understands itself as the liberation from that political domination that maintains itself through artificial preservation of the minority status of its subjects. Its politics are essentially pedagogic. It attempts to liberate the consciousness of humankind from the thought patterns of an imposed minority status, to teach men the use of their own insight. To the extent that equality is based upon common insight, the Enlightenment has egalitarian tendencies. To the extent that those few already emancipated want to extend the emancipation, it compels the formation of a temporary elitist structure, the avant-garde, a vanguard of progress.

This political Enlightenment takes part in the ambivalence of progress, easily detectable in the transition in emphasis from self-enlightenment to the enlightenment of others. The claim to be enlightened oneself is often nothing but naiveté. The claim that one is emancipated means frequently that one is actually not changing and is stuck at the stage of projecting rebellion, a preliminary stage of Enlightenment. In this book, investigating the reasons of ambivalence, we must always make the apparently opposite movement. We must strive for self-enlightenment with the question of what has been repressed in each instance in the movement that calls itself the Enlightenment. We proceed in this manner to investigate how humankind, in the course of history, has been asked questions concerning the dominant truth.

We proceeded from the problem of peace, and we cannot hope to find the criteria of the truth being sought without examining the international political problems of our time. Historical anthropology is always anthropology in the real history, that of the respectively contemporary world.

10.7 World Peace and Self-realization

What must happen in the contemporary world, the world in transition to tomorrow? What has to happen? Avoiding the obvious formulation of the question, "What should we do?" Who could be meant by 'we'? The weakness of the formulation, "What should we do?" is that it gives the impression that one has only to say what must be done and then do it. But as we have seen, this would be inadequate for two reasons. First, because this is, in this simple interpretation, itself a formulation from the world of will and insight. To attempt to answer this question in a simple way encourages too easily that behaviour that caused the need and perpetuates it. Basically no one reaches the point where he can act meaningfully without having gone through a period of despair about direct action. Secondly, any action that is potentially meaningful requires a complete rethinking of all questions. Much to my own dismay, I have often failed to reach in my own thinking the point where I could suggest the best course of action or what I myself should do. One answer I would give to the question, "What should we do?" would be: 'Think!' This was one of the reasons I founded an institute instead of going into politics.

The question "What has to happen?" formulates specifically an intellectual task. Action without theory takes place anyway. We do not have further to encourage it. But what should a guiding theory of well-thought-out action look like?

In my opinion, the guiding political thought has to be world peace. It is the only thought that can integrate worldwide political activity. It is obviously necessary for the avoidance of a catastrophic Third World War and, if this cannot be avoided, for the prevention of the next one. This necessity cannot be denied, even if it fails for the time being. On the other hand, much is demanded, and it is a hard criterion with which to judge all other political thoughts. Finally, this guiding principle can be understood by everyone even if its consequences are not easily comprehended by everyone. Peace should not be tyrannical. It has to embody a truth. Thinking about this truth is the meaning of our political theory.

I consider the happiness-oriented post-industrial society incapable of mastering the ambivalence of the systems that it will tolerate in order not to be disturbed in its happiness. If it is unavoidable that the most important value for the majority of humankind is subjectively experienced personal happiness, then this fact compels an elitist structure of society. The elite, if we are lucky, are at least partially oriented towards that truth that embodies world peace as a guiding political value. For the moment, this structure exists anyway to the extent that only a minority realize the existing problems, then assume a pedagogical function for the others.

What is the essence of the values in whose context lies the truth that inspires world peace? In the course of our reflections we have come upon several key words in trying to define these values. The meaning of these words is important as well as the context of these meanings. I have divided some of these words—gathered rather incidentally—into three groups. The first group could be freedom and social justice, the second group could consist of words like place, prosperity, physical and psychological health, love, partnership, and cultural creativity, and the third group could be insight, progress, maturity, and self-realization. I have explained the contents of each of these words at the appropriate place. What is their context, their relationship to world peace, and, consequently, their political relevance?

The list is framed with the concepts of freedom and self-realization. They can be considered the most important words of its respective groups. What do they mean? Freedom is essentially the freedom of self-realization, for only self-realization is actually freedom. Therefore, all depends on what is meant by self-realization. Happiness is an indication of realized values. Suffering may provide the impetus. The indicator is not always reliable. We differentiate between true happiness and questionable, partial, compensatory, false happiness. I believe that happiness is ultimately an indication of the stages of self-realization. This realization occurs in stages, therefore terms like progress and maturity also appear here. What was happiness at earlier stages is a means at higher stages, or it has to be sacrificed in order to reach the higher stage. The ultimately decisive knowledge about man is the distinction between the ego, with its need for happiness, and the self. The ego, in all its domains, is one of the stages leading to the self. Remaining at this stage is the origin of ambivalence. Whoever does not succeed in sacrificing this persistence, succeeds neither in self-realization of person, nor in the community of society. Accordingly, this sacrifice, and the realization made possible through it, and the resulting happiness, can only occur in stages, in the individual and, perhaps, in history.

This self-realization has been the theme of religion through the entire history of humankind. In concrete historical terms, it was the theme of religions in their plurality. The social embodiment of the achieved realization was cultures. Each such culture was a relative peace of a truth of human self-realization. World peace, if it is possible, is the peace of cultures encountering each other. However, the culture that compels it now, the Western culture, compels this through an uncontrolled progress of knowledge and power that has not obtained the goal of a recognized truth and is itself essentially without peace. This culture in its present form is an ambivalent secularization of its own religion. World peace might be compelled, for the time being through political events, as the external coexistence of cultures. But our own culture in its present state cannot even exist with itself, let alone coexist with others. As paradoxical as this may sound from the viewpoint of a strictly political or social analysis, I am convinced that the actual process through which the possibility of a true international peace can take place is dependent upon the realization of the human self. One of its forms is the entering of religion into its truth, which may be assisted by a meeting of cultures and religions. Entering into truth means realization of what is actually meant here, maturing. Political efforts are a part and a consequence of this effort for realization.

Humankind understands very well the demand for the sacrifice of the ego. In respect to political values, this is actually everywhere their core. For that side of the values that confirms the ego in its wishes does not need a demanding emphasis. The unity of absolutism results from forced sacrifice of particular interests. The freedom of liberalism is essentially the freedom that I grant my fellow citizen. The solidarity of socialism always demands the sacrifice of the personal in the name of common interests. There is perversion in all these values in that the ego converts them into demands made on others rather than on itself. The insight of these political values is usually limited by replacing the egoism of the individual with the egoism of the group, in which the undifferentiated values of the immature ego merely reproduce themselves. The task is to recognize and describe the true relation of self-realization to the human community of world peace.

Carl Friedrich von Weizsäcker 1992 at his "Alm" in the Alps. © Lili Bansa who granted permission to use this photo

Chapter 11
On Power

11.1 The School of Political Realism

The historical function of the school of political realism was to correct American historical optimism in reference to the realities of power politics.[1] I believe it has essentially contributed to the possibility of present American world-power politics.[2] The traditional American self-interpretation held that power politics (originally the power politics of monarchies) was historically a European evil in which the immigrants in the New World, from the Pilgrim fathers to the Germans fleeing in the 1848 revolution to the Jews fleeing Hitler, refused to participate. This view left the Americans with only two attitudes toward foreign policy that were consistent with their moral self-respect (and lively self-criticism): either isolationism, or interventionism for the purpose of restoring freedom and justice. Thus came about either renunciation of power politics (outside of America) or power politics aimed at breaking the power of power politics in the world. Neither attitude has been compatible with actual international politics of the nation USA as practiced over many decades. Therefore I maintain that no other imperialist power has tolerated to be forced into the imperialist role as much as America against its own intention and tradition. The discrepancy between moral self-interpretation and imperial reality has led to many inconsistencies in American politics. At times these have been likable, at times irritating, at times terrible. Knowing that one is morally on the right side of the war to end all wars means legitimization of actions that, under any other point of view (for example, as the actions of another nation), could only be called criminal.

Political realism is, in this respect, a basic introduction to politics for Americans. It teaches them to understand how power politics functions everywhere, even where they themselves participate in it. This lesson, especially with Niebuhr, also

---

[1] This text was first published in: *The Ambivalence of Progress* (New York: Paragon House, 1988), pp. 161–171, translation of: *Der Garten des Menschlichen* (Munich: Hanser, 1977): II.5.

[2] I am deeply indebted to G.-K. Kindermann: *Hans J. Morgenthau und die theoretischen Grundlagen des politischen Realismus. Einführung zur deutschen Übersetzung von* "Politics Among Nations" (Gütersloh: Bertelsmann, 1963): 19–47.

M. Drieschner (ed.), *Carl Friedrich von Weizsäcker: Major Texts in Philosophy*, SpringerBriefs on Pioneers in Science and Practice 23,
DOI: 10.1007/978-3-319-03671-7_11,

with Kennan, contained a strong moral impulse: These Americans, who were wise in the ways of the world, realized the extent of the involvement of Americans in objectively immoral activities in a world whose functional laws they did not understand. I consider this an inevitable process of shaping political consciousness in this specific historical context.

But the question must be asked as to whether this demand for an anthropological foundation has not meant that historically relative incomplete insights into politics have become dogma and simultaneously an ideological function that transformed the American nation, which once naively believed it could overcome power politics, into a world power as all the others. I admit that power politics, by and large, functions in the way described by political realism, but I doubt methodologically whether this can be attributed to anthropological reasons.

First, though, we will discuss the question of what is understood here by anthropology.

It is no coincidence that the foremost thoughts are those of a Protestant theologian, namely, of Reinhold Niebuhr. Kindermann (p. 22) quotes St. Paul: "The good thing I want to do, I never do; the evil thing which I do not want—that is what I do." (Rom. 7:19) A basic human experience is expressed here with a precision that is almost uniquely found in Pauline Christianity. In the light of this thought, humans are initially divided into those who are naïve or ideologists who manage not to know this about themselves, and those who do know it. Each of us actually incorporates both sides. We will later discuss the anthropological theories that see this dilemma itself as a result of history and that hope to overcome it through history.

The ancient Christian myth of history also belongs to these theories, because it traces humankind's sinfulness back to Adam's fall, and believes that it will be historically overcome on the Day of Judgment. To this extent the myth is not exactly 'anthropological' in a strict sense of scientific laws. But, actually, it is not science at all. Ever since the emergence of Christian dogmatics, meaning the self-interpretation of Christianity through the concepts of Greek philosophy, it has been under pressure to prove itself rational. I assume that all of these attempts at rationalization (including my own), as necessary as they are, remain inferior to the experience of reality transformed into myth, and thus distort it. We must now, though, follow the direction of modern scientific interpretation of this ancient experience.

At its starting point, the anthropology of Niebuhr and Morgenthau is neither naturalistic nor dualistic; it does not derive humans from nature, nor does it understand the tension of evil as the tension between nature and mind. It finds contradiction in the mind itself, namely, in the nature of freedom. Accordingly, it stands in a great philosophical tradition. But the proof that freedom by its nature has to lead to the known historical tragedies of power is not possible without intervention by a more empirical anthropology. In this context, Morgenthau develops a psychology of instinct that presupposes a division of the instincts into those of self-preservation, procreation, and power; which at first sight seems to presuppose what it is supposed to prove. Thus, he has to explain the emergence of the power instinct itself. After reading Kindermann's lecture, I feel that the obscurities of a causal-genetic anthropology, with which we have to deal later, are as little clarified here as in other existing anthropological statements. We are probably on safer ground again with the phenomenological description of what power means in contemporary international politics.

According to Kindermann (1963: 26), realists see power as becoming "a fundamental concept of the science of politics", comparable to the role of the concept of energy in physics (according to Bertrand Russell). A very interesting philosophical statement has been made here. I believe that mass, energy, information, and power, if understood with sufficient abstraction, are, in fact, identical in essence.[3] The statement articulating the identity of these different concepts only makes sense if their roles can be articulated. An attempt to do so is this verbal definition by Max Weber: "Power means any chance of imposing one's own will within a social relationship, even against resistance, no matter upon what this power is based." Here the concepts of society ('social relationship'), will, resistance, and chance (thus possibility) are assumed as understood. The question now is whether this concept of power is sufficient to substantiate the "typologically understood characteristics of political power" (Weber as quoted by Kindermann), which I can condense as follows: (1) omnipresence in politics, (2) omnipresence of its abuse, (3) potential limitlessness of its drive for expansion, (4) a fundamental orientation towards gaining security, domination, and prestige, and (5) a tendency to serve its own purposes and simultaneously to disguise this ideologically.

If there are mainly conflicting interests of individuals or groups and if politics is defined as the arena where such conflicts are settled, then power is indeed omnipresent in politics. Securing power is then essentially an interest, because it is a means for the realization of interests. Accordingly, the characteristics of (1) and to a certain extent (2) are explicable. The other three characteristics, unlimited drive for expansion, a tendency for serving its own purpose, and abuse and disguise, presuppose characteristics of humankind that we know only too well from experience, but whose origin and necessity cannot be made comprehensible through concepts like society, will, resistance, and interest. Here we must use as explanation something more relevant than these last-mentioned and somewhat vague generalities. The open question is, "Are these continuing characteristics anthropologically necessary? If yes, why? If no, how can they be historically explained?"

And if we thus inquire historically, we must also consider how valid, in general, are the extremely 'demonic' characteristics of power. The authors themselves have been marked by a certain historical experience. They are contemporaries of Stalin, Hitler, and the atom bomb; World War I was the horizon of their youth, a Third

[3] Cf. "Matter, Energy, Information", reprinted as Chap. 11 in: Michael Drieschner (ed.): *Carl Friedrich von Weizsäcker: Major Texts in Physics* (Cham et al.: Springer-Verlag, 2014) and originally published in: *The Unity of Nature* (New York: Farrar Straus Giroux, 1980), pp. 274–294; it was translated by Francis J. Zucker from: *Die Einheit der Natur* (Munich: Hanser, 1971): III.5.

World War is on the horizon of the future. Is power in history always like this? They speak of a tendency that must be continually resisted. Politics becomes immoral only where it idealizes the factual political behaviour of humanity into a norm. Is not successful resistance to this tendency the actual normal for humanity? Does the Pauline consciousness of sin understand humankind more clearly, or is it itself pathological?

The dispute about the relative importance of a tendency can hardly be decided empirically. The phenomenological position emphasizes, out of this multitude of events, a structure that is perhaps the decisive one for our own inner attitude towards history. For a more objective insight we can attempt at least to understand how it could, and perhaps had to, result in such a structure or tendency. So we take the detour of science.

11.2 Biological Anthropology

11.2.1 Evolutionist Thinking

First we must make clear the necessary level of the question.

A question that is still naïve though meaningful: Is a power instinct part of human nature? Initially one may try to determine the characteristics of this assumed power instinct. Further one has to ask, "What is understood by the word instinct: a vague desire or a structured scheme of behaviour, the latter possessing an innate structure or an innate capacity to be structured or imprinted by experience?" This then leads to the question of what one wants to understand as human nature. Is it an innate system of patterns of behaviour, or perhaps only the ability to acquire such patterns of behaviour? Is it in the nature of humans to have a history?

What appears to opponents of biological anthropology as a biological error, namely, the establishment of an invariable 'human nature' comparable to the nature of an animal species such as the nature of the wolf, is specifically already an inadequate biology. Compared with this, the necessary standard for the formulation of the question is to be found in the theory of evolution. According to it, the nature of the wolf is a stage in a historical process that is incomprehensible without knowledge of its ancestry, its ecological niche, the environmental changes caused by it, the reaction of its behaviour to these changes, and so on. Humankind's manner of having a history is very different from that of animals. We can, however, conceptually label this difference only if we have understood how animals, life itself, and nature can have histories.

The constant evolution of more and more differentiated forms is not a specialty of organic life. It only proceeds much further there than in inorganic nature. The evolution of new forms results from exactly the same probability considerations that also result in the second law of thermodynamics. Both are consequences of the same structure of time. Entropy and information are, according to Shannon, identical. Information is, again according to Shannon, a measure of the expectation value of the knowledge one could gain. Entropy is a measure of knowledge, similarly structured and understood, that one could have if one knew the complete microstructure instead of just the macrostructure. With a suitable form of interacting forces and sufficiently low temperatures, a state of high entropy will at the same time have a great abundance of forms. The thermal death of a closed system is then not a mush, it is many skeletons. The irreversible process far from equilibrium contains in such cases the continual increase in an abundance of moving forms. All of this is a consequence of the structure of time, as the condition of the existence of something like nature.

The organisms are distinguished from the forms of inorganic life by self-reproduction, mutation, and selection. The decisive achievement here is self-reproduction. Mutation is originally only a margin around precise self-reproduction. Selection is the consequence of the struggle for survival that is itself a consequence of the abundant production of forms through reproduction. Selection eliminates almost all mutants and therefore is the most important stabilizer of the species. It also favours the rare advantageous mutation through the additional influences of spatial isolation, temporal variations of environmental conditions, and so on, as described in the theory of selection. Selection favours not only particular types, but also type-forming tendencies, for example, those developing faster, such as the ability to store recessive characteristics or the relatively short life span of each generation.

There is no proof that the selection theory is sufficient to explain the evolution of organisms, and proof need not be assumed here. It suffices for now to see that there is selection pressure in favour of certain structures. We then have to examine, case by case, the relevance of such observations.

11.2.2 The Structure of Instincts in Humans and Animals

The usual way of speaking about human instincts is hybrid or at least thought-provoking. Let us take as an example the three basic instincts of political realism that I quoted above: self-preservation, procreation, and power. Here instinct is apparently understood as an irrational drive causing certain kinds of behaviour whose purpose can nevertheless be rationally articulated. The instinct is irrational insofar as it operates without insight into its purpose and even can assert itself against the insightful decisions of the will. Instincts can, de facto, shake the very structure of rational purposeful behaviour, or even make it collapse. This is, after all, one of the fundamentals of the Pauline experience of sin. What justification exists then to label instincts in accordance with rationally formulated purposes? By what authority are these purposes set, and what determines whether certain instinctual actions actually fulfil the purpose of the instinct? Procreative instinct is basically a *petitio principii* when one realizes how far separate the sexual instinct is from the idea of procreation. However, with procreation and self-preservation, reflection can at least recognize a biological meaning, whereas with power even this is doubtful. Freud finds stages in the unconscious that partially operate as if they were conscious. In his theory, however, these unconscious stages have explanatory value as the causes of 'irrational' behaviour only because the rationality of the unconscious is different from that of the conscious. Freud's great early discovery is the comprehensible significance of unconscious actions. But the utter strangeness of this comprehensible significance, and its methods of operation, forces Freud to search further and further for causal or structural hypotheses to explain the basically blind instincts. His later theory of instincts reads like a gloomy myth whose deepest affect is the horror of the inexplicable within ourselves. It is a great poem of concepts that has a fascinating unverifiability. But however one might feel about it, it does make us realize how little we know about ourselves.

If we proceed from Freud's image of humanity to what behavioural scientists teach us about animal behaviour, it is as if we are emerging from night into moderately bright daylight of comprehensible details. It might be that the clinical cabalists perceived other phenomena than the optimistic natural scientists, but in any case the perceptions of the latter are open to rational discussion. Perhaps the most important discovery in this respect is that among animals there are not simply a few basic instincts; there is a highly complex web of many independently interlocking and differentiated patterns of behaviour, some of which can be experimentally made to fail individually, or to be deceived by decoys, or, through new environments, to be directed toward new biological purposes. An implicit basic assumption of Freudian psychology, one that is presumably valid for the repressed contents of human consciousness, namely, that the unconscious knows what it wants, is not applicable here (unless one wants to fall back upon panpsychistic metaphysics). It is obvious that the animal does not need to be aware that by eating, copulating, and breeding it is preserving itself and the species. The innate instinct can be described as being subjectively totally irrational and unknowing because it permits itself to be deceived in a manner that is easily recognizable to us but is completely senseless for biological purposes. At the same time, it has a rather admirable 'objective rationality'. Evidently this admirable aspect is so much taken for granted by scientists that they usually do not bother to express their fascination. I reminded Konrad Lorenz of Kant's remark about a bird's nest, "One can only kneel before it and adore it," Lorenz replied, "I feel exactly like that every day." This adoration is, however, not scientifically idle. It seeks the comprehensible causes of suitability and tries to find them in evolutionist thinking.

What can we learn from this about humankind? First we learn that an instinct can be highly differentiated, without a conscious or unconscious stage in individuals that understand it (with the exception of the still incomplete, specifically human stage of subsequent reflection). Comprehension of our own instincts does not lie in the past where the myth makers often see it. Perhaps it lies in the future. We also learn that especially the highly differentiated, once objectively purposeful instincts can lose their purpose merely through changing historical circumstances.

They are as subjectively irrational later as they were at inception. All this takes us far away from the fervour of Freudian (and also Jungian) conceptual mythology. It is a fundamental contribution of scientific sobriety, indispensable even to someone like myself who believes very much in the relevance of myths.

The dispute with the environmental theorist begins where the behavioural scientist emphasizes the innate and, to that extent unplastic, character of instinctual behaviour. The dispute, factually very interesting and relevant, and whose ideological component is evident, does not have to concern me in regard to the problem of power because I am entirely willing to define the human being through its capacity for tradition and insight, and because I understand power as a human attribute due to this capacity. In regard to a theory of insight (cybernetics of truth), the structural relationship of innate disposition to achievement that is not innate has to be studied, a disposition for which there are many simple examples in animal behaviour.

But the phenomenon of the convergence of innate and acquired patterns of behaviour is of the greatest importance. In evolution theory, convergence is the appearance of similar facilities in genetically independent organisms, for example, the eye lens in vertebrates and octopods. Certainly it is not by chance, given so many human achievements, that one can dispute whether they are innate or acquired. Such a dispute can only arise if innate and acquired behaviours resemble each other. This can be expected when similar causative factors are operating in both cases. The result is not only a matter of purposeful solution to the same problem, such as adaptation to certain environmental conditions (for example, the convergence of the streamline shape in fish, seals, and whales, as well as that of airplane bodies and submarine hulls). It is just as much a question of behaviour patterns that is without apparent purpose or even undesired by humans, but upon which the pressure of selection is operating. One will not understand the phenomenon of power without examining such convergences.

11.2.3 Stabilization in Inequality Through Selection

Free competition favours the stronger. There exists a continual pressure toward increasing inequality. By slightly shifting the meaning of the concept, one can talk about a selection pressure in favour of inequality. This shifts the concept: One talks about selection pressure in favour of a characteristic of an individual (for example, body height) if this characteristic favours the individual in competition. If the characteristic is inherited, then selection pressure will favour the survival of those in which this characteristic distinctly exists. This does not always mean a continuous development of the characteristic if this pressure is countered by opposing pressures. One can also talk about selection pressure in favour of a characteristic of entire societies if, due to this characteristic (for example, a higher fertility rate), they can assert themselves against competing societies. But inequality is not a characteristic but a reflective concept on a higher level.

Inequality rises within a society, for example, if the stronger members can differentiate themselves more from the weaker ones. In this regard inequality is then a characteristic of this society and not of the individual members. But it is not necessarily a characteristic that gives this society a selection advantage over other societies; it may be the opposite. Inequality thus is a consequence of selection without necessarily offering an advantage to an individual or a society.

The existence of selection pressure does not in general mean the process of an actual evolution in this direction. Because the conditions that we can observe, at least in animal life, are somewhat established equilibriums, it follows that the equilibrium has been shifted in the direction where the pressure has an effect as compared with the fictitious example of an equilibrium without this pressure. Thus the pressure toward inequality usually does not lead to growing inequalities but towards an equilibrium with established inequalities. In the ecological equilibrium of a species this means, for example, that big, small, and tiny fish live together. Within a society, it is the social hierarchy establishing itself. Where competition and equilibrium exist simultaneously, equality between competing individuals is just as 'unnatural' as the extermination of the weaker by the stronger.

Therefore my thesis that equality among humans is not something natural, but an accomplishment that needs a moral postulate to be realizable. With respect to economic causalities, it seems to me to be equally wrong to expect from a market economy the preservation or establishment of economic equality, or to assume, in contrast, that it must destroy itself through monopolistic concentrations. The former is not valid at all; and the latter only if one knows from elsewhere that there is no possibility for establishing an equilibrium. Regarding human competition, one has to consider the specific human phenomenon that serves competition and is the real central issue: power.

11.2.4 What Is Power?

I assumed above that power was identical with mass, energy, and information, an *assumption that requires not only the specification of the characteristic* genus proximum, *but* also that of the *differentia specifica*. Mass, energy, and information are quantities. There is more or less of each of them in comparable cases. But this is much less obvious when applied to power, though one can be more or less powerful. Specific economic power is, to a certain extent, measured by the quantity of money. But what is the essence of the thing being measured here?

In my text "Matter–Energy–Information"[4] I define mass as a measure of matter, energy as a measure of movement, information as a measure of form. Energy and information are as much the measure of potential as of actual movement and form. The mechanical law of the conservation of energy, for

[4] Cf. Footnote 3.

example, applies only to the totality of potential and kinetic energy. But how could one measure movement and form?

To the extent that reality presents itself to us as moving form it is actually an entirety that is stripped of its essential characteristics through every act of mental dissection. But where forms can be regarded as identical, they can be counted, and thereby the context that differentiates them is lost. It is this 'regard as identical' that is the basis of the formation of a concept. Physics is based on this counting. Determining whether in a given situation a certain form does or does not appear means deciding on an alternative. A part of the entirety of reality that is lost by dissection into alternatives is found again in the form of the laws that dominate the choice between alternatives, laws that we call the laws of nature. Physics is accordingly a partial mental restoration of the entirety that has been divided into alternatives through the dissection of the real—which makes physics possible. The total of possible independent decisions on alternatives is information. The capability to make such decisions can be called power. In this regard one can state that the information available to someone is a measure of the individual's power.

There is form, and consequently information, on very different levels of integration, on very different semantic levels or under different concepts. Information on the level of the most extreme dissection we can achieve constitutes, if I interpret physics correctly, matter and energy as defined by physics. The power of a human individual or a group can be measured by the information available to them. This information does not affect the lower levels of integration. The power to dissolve these is usually only destructive. One can attempt to understand power as information on a level of integration specific to humankind. This, however, can and must be elucidated.

Recognition of the parallel between information and power turns a first, dim light on the cause leading to the possibility of the egocentricity of power as observed by political realism. I have said earlier that information is only what produces information. In which sense does one use these words if one formally translates this into the assertion that power is only what produces power?

I can only indicate the not entirely obvious and deeply structured levels of abstraction behind this interpretation of information. The philosophical background is a philosophy of time that understands form essentially as motion and, in this regard, as self-renewing. Against this background stands the question of the way information relates to the recipient. To whom is it information? Whose concepts are controlling its definition in each case? Biologists have to use the concept of information objectively; the genetic information of a DNA chain shows its semantic ability in the produced phenotype. Here one has to differentiate between actual and potential information. Shannon's information as expectation value $H = \sum_k p_k \cdot \log p_k$ of the news value $J_k = \log p_k$ of the possible event k is potential information; J_k, if k has occurred, is actual information. The thesis that information produces information contains then two very different laws in justification and application, a law of conservation and a law of growth, parallel to the two principal laws of thermodynamics. The first is connected with the operative-objective definition of information and cannot be further pursued here. The latter is a thesis of evolution: Potential information increases with time.

The thesis of evolution can now be transferred with crude selectivity to the concept of power as commonly understood. Power is only what produces power. In any power competition, the one who can accumulate power will prevail. This is the abstract core of Marxist analysis of capitalism. Our reflections show that this fact does not depend on power having here the form of economic capital. Therefore I maintain that the abolition of capitalism does not, as such, liberate us from the problem of power. If we hope ever to overcome power, then we have to abolish the 'crude selectivity' structure not only in our opinions but in fact.

In which sense can I actually call power a humanum, something specifically human, if I connect it with energy and information and continually argue in the context of biological evolutionism? In the last chapter[5] I have described power as a product of insight, of isolated or isolating insight that I understood as a humanum. Now I have introduced information within the framework of exactly the same kind of insight as the quantity of possible alternatives. Accordingly, I have introduced information itself as a measure of power. Thus information itself is a humanum: Only to human thinking in power categories, nature is a structure of information. But energy, it seems to me, is information, and matter is energy. Physics and biophysics are then themselves ways of thinking about nature in power categories. A physically understood nature itself is a humanum. This is not a new thought. We find it in Heidegger, and in Marcuse, who probably took it from him. But in its effect it remains destructive or romantic as long as we cannot limit the scope of the validity of power thinking, employing deeper-rooted premises. There we have to reckon with a great challenge if it is correct that the isolation of recurring forms characterizes not only quantitative physics but conceptual thinking.

In summary, the subsumption of human power under information itself belongs to the method of thinking in categories of disposition or power; on the other hand, within the frame of this subsumption, power is a specific kind of information occurring in humankind and connected with the accumulation of tradition and insight.

In the chapter "The Rationality of Emotions"[6] I connected power and peace with insight without specifically discussing their reciprocal relationship. I connect power there with the limitlessness of isolating knowledge; later this realm appears under the name of volition and intelligence. I define peace as the embodiment of truth; this corresponds later to the name of reason. Among the concepts used here, intelligence must be employed as the capacity for an isolating kind of knowledge, and reason as the capacity for understanding an entirety as an entirety. Intelligence is unlimited because it omits the natural boundary of a part within a whole. On the other hand, there is no total antagonism between intelligence and reason, only one that is supplemental or rather subordinate. To isolate is itself an achievement that,

[5] Chapter 10 in this volume—*editor* [MD].

[6] ibid.

like any achievement, presupposes an entirety it can achieve. This accomplishment usually occurs without insight into its own nature ("Consciousness is an unconscious act"). If it is recognized, however, it is an act of reason. Intelligence thus presupposes de facto something that is accessible only to reason. The achievement of intelligence does not consist of dissecting. Intelligence recognizes the context of what it has isolated in laws. In physics, too, the method of formulating the problem is power, the knowledge of the law is peace.

In answer to the question of whether these thoughts have sought to establish power as a fundamental biological-anthropological concept, I would say, "Not empirically, but in a certain sense transcendentally." "Not empirically" means I do not need to assume that there exists an innate power instinct comparable to the innate sexual instinct. But neither do I have to contest such an assumption. "In a certain sense transcendentally" means that power appears as the unavoidable result of the coincidence of any competitive situation and intellectual understanding. In view of the possibility of reason and peace, power is not necessarily the last word.

That is for history to decide.

Carl Friedrich von Weizsäcker Society

WISSEN UND VERANTWORTUNG
Carl Friedrich von Weizsäcker-Gesellschaft e.V.

Knowledge and Responsibility
Carl Friedrich von Weizsäcker Society

Modern science, especially the natural sciences, has given us the power of Greek gods. However, we would need the wisdom of Solomon to use this power sensibly. This is not something we have achieved, but rather it is a task facing us—possibly the single most important task of our time. In 1994, the Carl Friedrich von Weizsäcker Society had 18 founding members. Today, the Society's activities include the organization of international symposia and the development of projects on the decisive challenges of our time.

Knowledge Means Responsibility: Responsibility Needs Knowledge

"Knowledge and Responsibility" is our programme in a nutshell. Inspired by the concerns and by the work of Carl Friedrich von Weizsäcker, the Society tries

- to achieve an unbiased and rigorous analysis of our time in five working areas, and
- to develop projects that particularly address the challenges and responsibilities of our time.

"What must we do?" is first and foremost a question of insight; but it carries with it the task of furthering insights by gaining them a hearing and weight. Key programme tasks of the *Carl Friedrich von Weizsäcker Society* are therefore, for example, public conferences, expansion of membership, sponsors, partners and friends; but also to strive to develop in the longer term a "network of reason".

M. Drieschner (ed.), *Carl Friedrich von Weizsäcker: Major Texts in Philosophy*, SpringerBriefs on Pioneers in Science and Practice 23,
DOI: 10.1007/978-3-319-03671-7, © The Author(s) 2014

Areas of Activities

Physics, philosophy, theology, economics and *altered awareness* are the areas of activity that will be addressed in our projects. Throughout his life as a scholar, Carl Friedrich von Weizsäcker has continued to address these areas. This is one motive for your choice. The second is the way they create our history and our future: nowadays all societies and cultures more or less depend on scientific and technical civilization, up to and including the solution of their economic and social problems. Still, *physics* may be considered as a "key science", *philosophy* as a warning voice, "Do you know what you are saying, and do you know what you are doing?" *Theology* is the effort to understand what religious tradition can teach us for today and tomorrow, *economics* tries to understand social, environmental and political problems. *Altered awareness*, finally, the fifth area of activity, and which pervades all the others, explicitly or implicitly, systematically addresses questions of action and ethical stance in our time.

Address Prof. Dr. Thomas Görnitz (chairman), Dr. Bruno Redeker (executive chairman), Carl Friedrich von Weizsäcker-Gesellschaft e.V., Bielefelder Straße 8, 32130 Enger, Germany.
Website www.CFvW.de and www.CFvW.org.
E-Mail wuv@cfvw.de.

Carl Friedrich von Weizsäcker Foundation

CARL FRIEDRICH VON
WEIZSÄCKER-STIFTUNG

CARL FRIEDRICH VON WEIZSÄCKER FOUNDATION

Modern science, especially the natural sciences, has given us the power of Greek gods. However, we would need the wisdom of Solomon to use the power sensibly. This is not something we have achieved, but rather it is a task facing us—possibly the single most important task of our time. The activities of the *Carl Friedrich von Weizsäcker Foundation*, established in 2002, focus on the organization of international symposia, on the preservation and publication of the scientific legacy of Carl Friedrich von Weizsäcker, and on the development of projects on the key challenges of our times.

The Central Guiding Questions

What should we know? What must we do? What may we hope for?

Immediately bring to mind Kant's "What can I know? What should I do? What may I hope for?" At the same time they imply a change of perspective towards reason jointly applied to the challenges of our time, the practical problems that humankind faces today:

- Science and technology model a world without borders,
- Innovations, technology and the market drive change in our time,
- The global population is growing and increasingly divided into 'young' and 'old' societies,
- The gap between poverty and wealth widens ever further: locally, regionally and globally,

M. Drieschner (ed.), *Carl Friedrich von Weizsäcker: Major Texts in Philosophy*, SpringerBriefs on Pioneers in Science and Practice 23,
DOI: 10.1007/978-3-319-03671-7, © The Author(s) 2014

- The potential for war and terrorism continues to grow, encompassing ethnically and culturally driven conflicts,
- Our use of resources is increasing, placing stress on the biosphere,
- Human power challenges the inherited constitution of nature,
- Overall political order is dominated more and more by the laws of the market,
- Democratic influence on political processes and decisions is waning,
- Ethical stances become relative in the bazaar of opinions.

In the Chap. 8 of his book *Der Mensch in seiner Geschichte* [Humankind in its History] von Weizsäcker reflects on his adaptation of Kant's questions under the heading "Where are we going?": poverty and wealth, war and peace, human beings and nature, the problems are not resolved. But "with jointly applied reason they would be solvable". This is what Carl Friedrich von Weizsäcker has argued for throughout his life as a scholar: not from the perspective of a developed theoretical system but with rationality following the example of everyday speech, "Be reasonable!" And "Our task for today is the global search for truth". And "Reason means recognizing the necessary, and applied in common, to bringing into being what has been recognized as necessary." If we fail to broaden and deepen our understanding of what lies at the core of the challenges of our time as far as we can, there is a constant danger that we might cause more harm than good. "Hope is the perception of the possible" wrote von Weizsäcker in answer to his third question, and at the end of his book he speaks of his hope in these words: "I have tried to speak about what I have experienced. Others may experience other things, more things. They will act."

Address Dr. Bruno Redeker (chairman), Carl Friedrich von Weizsäcker Stiftung, Bielefelder Straße 8, 32130 Enger, Germany.
Website www.CFvW.de and www.CFvW.org.
E-Mail stiftung@cfvw.de.

Federation of German Scientists

FGS Federation of German Scientists

Founding members: G. Burkhardt, C. F. v. Weizsäcker. W. Gerlach

The Federation of German Scientists (FGS; German acronym VDW) was founded in 1959 in West-Berlin by renowned nuclear scientists, including Carl Friedrich von Weizsäcker and the Nobel Prize laureates Max Born, Otto Hahn, Werner Heisenberg, and Max von Laue.

Two years earlier this group of experts had become well-known to the public as 'Göttinger 18': Nuclear scientists who had publicly argued against a nuclear armament of the German Bundeswehr. Since then the FGS feels bound to the tradition of responsible science. It has nearly 400 members from different fields of the natural sciences, the humanities, and social sciences, so that a large range of topics is approached at a high level of competence. With the results of its interdisciplinary work the Federation of German Scientists not only addresses the general public, but also the decision-makers at all levels of politics and society.

The members of FGS stand in this tradition. They feel committed to taking into consideration the possible military, political, economic and social implications and possibilities of atomic misuse when carrying out their scientific research and teaching.

In Annual Conferences and in interdisciplinary Expert Groups as well as public comments it addresses issues of science and technology on the one hand, and peace and security policy on the other. At the same time, the role of science itself in genesis and in solution of socio-technological problems is subject of examination and expertise. FGS' membership lists also include representatives of the humanities and social sciences, so that a large range of topics is approached at a high level of competence. With the results of its interdisciplinary work the Federation of German Scientists not only addresses the general public, but also the

M. Drieschner (ed.), *Carl Friedrich von Weizsäcker: Major Texts in Philosophy*, SpringerBriefs on Pioneers in Science and Practice 23,
DOI: 10.1007/978-3-319-03671-7, © The Author(s) 2014

decision-makers at all levels of politics and society. According to its statutes of 1959, the FGS aims to

- keep up and deepen the awareness of those working in science for their responsibility for the effects which their work has on society;
- study the problems which result from the continuous development of science and technology;
- assist science and its representatives in making public the questions related to the application of scientific and technical developments;
- provide advice and thus exercise influence on decisions as long as they are assessable and can be dealt with by means of scientific knowledge and methods, and to point out all forms of misuse of scientific and technical results;
- to defend the freedom of scientific research and the free exchange of its results and to expand and strengthen the traditional international cooperation of scientists.

Carl Friedrich von Weizsäcker had been part of the famous 'Göttinger 18', the group of renowned nuclear scientists who publicly opposed a possible nuclear armament of West Germany in the 1950s, and was among the founding members of the Federation of German Scientists in 1959. His spirit, his way of perceiving the world and his understanding of the role and responsibility of science for society and the development of humankind profoundly shaped the self-perception and sphere of influence of the FGS in its early years and later on. He also repeatedly served in public functions of the FGS, most notably as its chairman from 1969 to 1973.

Address Vereinigung Deutscher Wissenschaftler (VDW), Marienstr. 19/20, 10117 Berlin, Germany.
Email info@vdw-ev.de.
Website http://www.vdw-ev.de/index.php/de-DE.

VDW Vereinigung Deutscher Wissenschaftler

Udo Keller Stiftung Forum Humanum

UDO KELLER STIFTUNG
FORUM HUMANUM

Mission Statement

The name reflects the programme of action. The Foundation, set up by the Hamburg businessman Udo Keller, sees itself as a *Forum Humanum*—a forum for all those who would like to investigate the question of the truly human. At a time when technology and economic processes are increasingly influencing human choices, the Foundation addresses the importance of the moral and religious heritage of human cultures worldwide. The Foundation assumes that the future development of human beings will decisively depend on whether we succeed in harnessing the rich potential of these traditions for the future. In this way the Udo Keller Foundation argues for a revival of the question of the purpose of human life in twenty first century terms.

Funding Priorities

The *Udo Keller Foundation Forum Humanum* contributes to an interdisciplinary dialogue between natural sciences and the humanities as well as to the multi-faith dialogue between world religions. These goals are being realized at its

M. Drieschner (ed.), *Carl Friedrich von Weizsäcker: Major Texts in Philosophy*,
SpringerBriefs on Pioneers in Science and Practice 23,
DOI: 10.1007/978-3-319-03671-7,

headquarters in Neversdorf near Hamburg and at its study centre in Tübingen, the *FORUM SCIENTIARUM at the Eberhard Karls University of Tübingen.*

Funding Activity

The *Udo Keller Foundation Forum Humanum* is a co-founder of the interdisciplinary project *FORUM SCIENTIARUM at the Eberhard Karls University of Tübingen* and is one of several inaugurators of the *Academy of World Religions* at the University of Hamburg. The Foundation has sponsored the *Verlag der Weltreligionen* (World Religions Press) since its establishment in 2007, and has initiated various lecture series in Hamburg and Tübingen—including *Thinking the future* (ZUKUNFT denken) in Hamburg in cooperation with the Hamburg Planetarium (2010–2014) and the *Unseld Lectures* at Tübingen (from 2008). Together with the German Literary Archives in Marbach, the Foundation has funded since 2008 the *Udo Keller Scholarship for Contemporary Research into Religion and the Modern Age.*

Additional information on the work of the *Udo Keller Foundation Forum Humanum* may be accessed in German on its website at: www.forum-humanum.org.

Address Udo Keller Stiftung Forum Humanum, Kleine Seestr. 24, 23816 Neversdorf, Germany.
Email info@forum-humanum.org.

Ruhr University Bochum

RUHR
UNIVERSITÄT
BOCHUM
RUB

Portrait

Located in the midst of the dynamic, hospitable metropolitan area of the Ruhr, in the heart of Europe, the Ruhr-University Bochum (RUB) with its 20 faculties, RUB's disciplinary institutional units, is home to 5,600 employees and over 41,000 students from 130 countries. All the great scientific disciplines are united on one compact campus.

The RUB is on its way to becoming one of the leading European universities of the twenty first century. Almost all courses are offered as Bachelor and Master degree programmes. Our excellence programmes have made themselves an international name: Our Research School is an international college for structured doctoral research in the life sciences, natural sciences, engineering, the humanities and social sciences. Interfaculty and interdisciplinary Research Departments, which are mutually, nationally and internationally networked, sharpen the profile of the RUB. Added to this is an unsurpassed programme for the promotion of Early Career Researchers, and an excellent infrastructure.

What makes it all come alive is the people who meet on campus with their thirst for knowledge, their curiosity, and their commitment. They help shape the RUB and their open-mindedness makes the RUB an attractive place for people from around the world.

M. Drieschner (ed.), *Carl Friedrich von Weizsäcker: Major Texts in Philosophy*,
SpringerBriefs on Pioneers in Science and Practice 23,
DOI: 10.1007/978-3-319-03671-7,

Guiding Principle of the RUB

The trio of values, people-centred—cosmopolitan—high-performance, represent the cornerstones of the RUB environment. This space is more than just the sum of its individual elements: People-centred and cosmopolitan means to respect diverse cultures and to give guests a home. People-centred and high-powered means jointly developing creative forces, to 'tackle' things with verve and ambition. 'Campus RUB' is the contemporary *universitas*—the community in which people take centre stage.

Living Universitas

The members of the *universitas* teach others and, at the same time, learn from each other—whether in science, studies, engineering, or management. *Universitas* promotes and demands codetermination from everyone—a far cry from academic or hierarchical structures. Living universitas is: When students in the eTutoring project plan courses; when the promotional funding line of the research fund is judged solely by doctoral students; the uncomplicated interaction of researchers from various disciplines.

Address Ruhr-Universität Bochum, Universitätsstraße 150, 44781 Bochum, Germany.
Website http://www.ruhr-uni-bochum.de/universitaet/index_en.html.
Email international@rub.de.

About the Author

Carl Friedrich Freiherr von Weizsäcker[1] (June 28, 1912—April 28, 2007) was a German physicist and philosopher. A member of the prominent Weizsäcker family, he was son of the diplomat Ernst von Weizsäcker, elder brother of the former German President Richard von Weizsäcker, father of the physicist and environmental researcher Ernst Ulrich von Weizsäcker, and father-in-law of the former General Secretary of the World Council of Churches Konrad Raiser.

Born in Kiel, he was raised in Stuttgart, Basel, and Copenhagen. From 1929 to 1933, Weizsäcker studied physics, mathematics and astronomy in Berlin, Göttingen and Leipzig supervised by and in cooperation with Werner Heisenberg and Niels Bohr, among others. The supervisor of his doctoral thesis was Friedrich Hund.

Weizsäcker made important discoveries in theoretical physics regarding the masses of atomic nuclei, energy production in stars from nuclear fusion processes, and on planetary formation in the early Solar System. During World War II he participated in the German program for developing nuclear energy and atomic bombs. In his later career, he focused on philosophical and ethical issues, and was awarded several international honours for his work in these areas.

Work on nuclear physics Weizsäcker's special interest as a young researcher was the physics of the atomic nucleus. Simultaneously with Hans Bethe he found a mechanism or pathway for the cyclic process of fusion in stars (Bethe-Weizsäcker process, published 1937–1939). This discovery should not be confused with his 1935 development of the Bethe-Weizsäcker formula, or Semi-Empirical Mass Formula (SEMF) for nuclear masses, again simultaneously with Hans Bethe.

Work on planetary formation In 1938, Weizsäcker developed a theory of the formation of the Solar System, based mainly on considerations of turbulent motion

[1] The copyright of the photograph on this page is with © Max-Planck-Gesellschaft/Filser who granted permission for its use in this volume.

M. Drieschner (ed.), *Carl Friedrich von Weizsäcker: Major Texts in Philosophy*, SpringerBriefs on Pioneers in Science and Practice 23,
DOI: 10.1007/978-3-319-03671-7, © The Author(s) 2014

of gases and dust. The theory also helped to explain the empirically observed regular pattern of increase in the diameters of the orbits of the planets of the Solar System, from inward to outward.

Work on atomic weapons As a theoretical physicist, Weizsäcker (and by his own estimate, 200 other physicists) had recognized immediately after nuclear fission had become known (by Otto Hahn) in 1938 that nuclear weapons could potentially be built. He discussed the upsetting implications in February 1939 with philosopher friend Georg Picht.

During World War II, Weizsäcker joined the German nuclear energy project, participating in efforts to construct an atomic bomb. For some time he had been hoping for political influence growing out of participation in a successful nuclear weapons project. In July 1940 he was co-author of a report to the Army on the possibility of 'energy production' from refined uranium. The report also predicted the possibility of using plutonium for the same purpose including the production of a new type of explosives. During summer 1942 Weizsäcker drafted a patent on a transportable "process to generate energy and neutrons by an explosion … e.g., a bomb", which was never filed. The draft was found in the 1990s in Moscow.

Historians have been divided as to whether Heisenberg and his team were sincerely trying to construct a nuclear weapon. In a 1957 interview with the German weekly Der Spiegel, Weizsäcker frankly admitted to the scientific ambitions of those years: "We wanted to know if chain reactions were possible. No matter what we would end up doing with our knowledge—we wanted to know." Weizsäcker said that they were spared the decision on building the bomb as they saw rather soon that the German war economy was unable to mobilize the necessary resources.

Weizsäcker worked later during the war as a professor in Strasbourg. The American capture of his laboratory and papers there in December 1944 revealed to the Western Allies that the Germans had not come close to developing a nuclear weapon.

Post-war career In 1946, Weizsäcker became director of the department for theoretical physics in the Max Planck Institute for Physics in Göttingen. Weizsäcker felt that the scientists who had developed the foundations of such powerful theories as that of the atomic nucleus, should take on the responsibility for the consequences. In 1957, it was mainly he who formulated the protest of the 'Göttinger 18', a group of prominent German physicists, against the idea that the West German armed forces should be equipped with tactical nuclear weapons. He suggested that West Germany should declare its definitive abdication of all kinds of nuclear weapons. From 1957 to 1969, Weizsäcker was professor of philosophy at the University of Hamburg. From 1970 to 1980, he was head of the Max Planck Institute for the Research on Living Conditions in the Modern World in Starnberg. He researched and published mainly on philosophy and foundations of physics, but also on the danger of nuclear war, which he thought underestimated by the public and the political establishment, on the conflict between the First World and the Third World, and the consequences of environmental degradation, and on the world as an interlocking whole ('Weltinnenpolitik'). In the 1970s he founded, together with the Indian philosopher Pandit Gopi Krishna, a research foundation 'for western sciences and eastern wisdom'. See also the website on this book on Carl Friedrich von Weizsäcker, at:<http://afes-press-books.de/html/SpringerBriefs_PSP_C.F.v._Weizsaecker.htm>.

After his retirement in 1980 he intensified his work on the conceptual foundations of physics and on philosophical issues. In the 1980s he invested much of his creative energy in the promotion of what was originally called a "Council for Peace". The movement resulted in the "World Convocation on Justice, Peace and the Integrity of Creation" in Seoul in 1990.

Weizsäcker developed the theory of ur-alternatives (archetypal objects), publicized first in his book *Die Einheit der Natur* (1971; English translation *The Unity of Nature* (1980)) and further developed through the 1990s. The theory axiomatically constructs quantum physics and uses it to discuss the foundation of a universal physics on the quantum mechanics of binary alternatives. Weizsäcker used his theory, a form of digital physics, to derive the 3-dimensionality of space. The program has not, so far, come to an end. In 2007, Weizsäcker died at the age of 94 in Starnberg, Germany.

Awards and honours Max Planck Medal (1957), Goethe Prize of the city of Frankfurt am Main (1958), Pour le Mérite for Science and Art (1961), Peace Prize of the German Book Trade (1963), Erasmus Prize of the city of Herdam (1969), Austrian Medal for Science and Art (1969), Grand Merit Cross with Star and Sash of the Federal Republic of Germany (1973) Ernst Hellmut Vits Prize of the University of Münster (1982), Heinrich Heine Prize of the city of Düsseldorf (1983), Sigmund Freud Prize for Scientific Prose (1988), Templeton Prize for "Progress in Religion" (1989), Theodor Heuss Prize "for his world-renowned, diverse and dedicated contributions to humanity themes: peace—justice—Integrity of Creation" (1989), Prix Arnold Reymond (University of Lausanne), Hanseatic Goethe Prize, Karl IV Prize of the City and University of Prague.

Honorary degrees: *Law*: Free University of Amsterdam, University of Alberta, University of Aberdeen; *Theology*: University of Tübingen, University of Basel; *Science*: Karl Marx University, Leipzig; *Philosophy*: Berlin Institute of Technology, University of Aachen.

Memberships Max Planck Society for the Advancement of Sciences, German Academy of Sciences Leopoldina, Göttingen Academy of Sciences, Saxon Academy of Sciences, Austrian Academy of Sciences, Bavarian Academy of Sciences, Bavarian Academy of Fine Arts, German Physical Society, Académie des Sciences Morales et Politiques, American Physical Society, Croatian Academy of Sciences and Arts, German Academy for Language and Literature, Joachim-Jungius Society of Science/Hamburg Academy of Sciences, Hamburg Institute for Human Sciences.

Among his major publications are Zum Weltbild der Physik (Leipzig 1946, 2002, 14th edition, renewed and with introduction by Holger Lyre) [*The World View of Physics* (London 1952)]; *Der begriffliche Aufbau der theoretischen Physik* (Lecture Notes 1946) (Stuttgart 2004); *Die Geschichte der Natur* (Göttingen 1948) [*History of Nature* (London 1951)]; *The Relevance of Science (*London–New York 1964); [*Die Tragweite der Wissenschaft* (Stuttgart 1990)]; *Die Einheit der Natur* (Munich 1971) [*The Unity of Nature* (New York 1980)]; *The Biological Basis of Religion and Genius*, Gopi Krishna (New York 1971), intro. by Carl Friedrich von Weizsäcker, which is half the book; *Wege in der Gefahr* (Munich 1976) [*The Politics of Peril* (New York 1978)]; *Der Garten des Menschlichen* (Munich

1977) [*The Ambivalence of Progress*, Essays on Historical Anthropology (New York 1988)]; *Deutlichkeit: Beiträge zu politischen und religiösen Gegenwartsfragen* (Munich 1978); *Der bedrohte Friede* (Munich 1981); *Wahrnehmung der Neuzeit* (Munich 1983); *Aufbau der Physik* (Munich 1985) [*The structure of physics* (Heidelberg 2006)]; *Die Zeit drängt* (Munich 1986); *Bewusstseinswandel* (Munich 1988); *Der Mensch in seiner Geschichte* (Munich 1991); *Zeit und Wissen* (Munich 1992); *Große Physiker* (Munich 1999).

These books of C. F. von Weizsäcker were published in English The history of nature, translated by Fred D. Wieck (Chicago: University of Chicago Press, 1949, 1966) translation of: *Die Geschichte der Natur* (Göttingen, 1948); *The world view of physics*, translated by Marjorie Grene (London: Routledge and K. Paul 1952), translation of: *Zum Weltbild der Physik* (Leipzig 1946, 2002); C. F. von Weizsäcker and J. Juilfs: *The rise of modern physics*, translated by Arnold J. Pomerans (New York: G. Braziller, 1957), translation of: *Physik der Gegenwart*; C. F. von Weizsäcker and J. Juilfs, also published as: *Contemporary physics*, Translated by Arnold J. Pomerans (London: Hutchinson's Scientific and Technical Publications, 1957); *Ethical and political problems of the atomic age* (London: SCM press; Burge memorial lecture 1958); *The relevance of science; creation and cosmogony* (London: Collins, 1964); *The spectrum of turbulence with large Reynolds numbers*, translated by Barbara Dickinson (Farnborough: Ministry of Technology, 1966); L. Castell, M. Drieschner, C. F. von Weizsäcker (Eds.): *Quantum theory and the structures of time and space*, I–VI (Munich: Hanser, 1974–1986); *Cooperation of Western nations in a coming world crisis* (Palo Alto, CA: Aspen Institute for Humanistic Studies, 1977); *The politics of peril: economics, society, and the prevention of war*, translated by Michael Shaw (New York: Seabury Press, 1978). Translation of: *Wege in der Gefahr* (Munich: Hanser, 1976); *The unity of nature*, translated by Francis J. Zucker. (New York: Farrar Straus Giroux, 1980), translation of: *Einheit der Natur*; *The ambivalence of progress: essays on historical anthropology* (New York: Paragon House, 1988). Translation of: *Der Garten des Menschlichen* (Munich: Hanser, 1977); Plaass, Peter: *Kant's theory of natural science; translation, analytic introduction, and commentary* by Alfred E. and Maria G. Miller; with an introductory essay by Carl Friedrich von Weizsäcker. (Dordrecht–Boston: Kluwer Academic, 1994); revised and enlarged by Thomas Görnitz and Holger Lyre (eds.): *The structure of physics* (Dordrecht: Springer, 2006), based on: *Aufbau der Physik* (Munich: Hanser, 1985). And the five volumes in this series of *Springer Briefs on Pioneers in Science and Practice* (Cham–Heidelberg–New York–Dordrecht–London: Springer, 2014, 2015): Ulrich Bartosch (ed.): Carl Friedrich von Weizsäcker: *Pioneer of Physics, Philosophy, Religion, Politics and Peace Research* (PSP 21); Michael Drieschner (ed.): *Carl Friedrich von Weizsäcker: Major Texts in Physics* (PSP-22); Michael Drieschner (ed.): *Carl Friedrich von Weizsäcker: Major Texts in Philosophy* (PSP-23); Konrad Raiser (ed.): *Carl Friedrich von Weizsäcker: Major Texts on Religion* (PSP-24). Ulrich Bartosch (ed.): *Carl Friedrich von Weizsäcker: Major Texts on Politics and Peace Research* (PSP 25). See also the website on this book on Carl Friedrich von Weizsäcker, at:<http://afes-press-books.de/html/SpringerBriefs_PSP_C.F.v._Weizsaecker.htm>.

About the Editor

Michael Drieschner (born 1939) is a professor em. of Philosophy of Nature at the University of Bochum, Germany. After passing the 'Diplom' exam in physics (Munich 1964) he obtained his PhD in philosophy at the University of Hamburg in 1968 in the research group of Carl Friedrich von Weizsäcker with a work on the axiomatic structure of quantum mechanics. From 1970 to 1978 he was a researcher at the 'Max-Planck-Institute for Research on the Conditions of Life in the Scientific-Technological World' in Starnberg, Germany, again collaborating with C. F. v. Weizsäcker. In 1979 he published his *Voraussage—Wahrscheinlichkeit—Objekt* (Heidelberg 1979), a treatise on the foundations of quantum mechanics. From 1986 to 2006 he taught 'Naturphilosophie' (philosophy of nature) at the University of Bochum.

Further books are (Titles translated into English): *Introduction to the Philosophy of Nature* (Darmstadt: 1981/91); *Carl Friedrich von Weizsäcker—an Introduction* (Hamburg 1992); *Modern Philosophy of Nature* (Paderborn 2002). With L. Castell and C. F. v. Weizsäcker he was a co-editor of the first two volumes of: *Quantum Theory and the Structures of Time and Space* (Munich: Hanser, 1975 and 1977); he edited the collected works of C. F. v. Weizsäcker on CD-ROM (Berlin 2011). His papers on physical themes in English language are: "Lattice Theory, Groups, and Space", in: L. Castell, M. Drieschner, C. F. v. Weizsäcker (Eds.): *Quantum Theory and the Structures of Time and Space* (Munich: Hanser, 1975): 55–69; "Is (Quantum) Logic Empirical?", in: *Journ. Philos. Logic*, 6 (1977): 415–423; "The Abstract Concept of Physical Object", in: L. Castell, M. Drieschner, C. F. v. Weizsäcker (Eds.): *Quantum Theory and the Structures of Time and Space 2* (Munich: Hanser, 1977): 20–31; "The Subject Matter of Quantum Mechanics", in: *International Journal of Theoretical Physics*, 31 (1992): 1615–1625; "The Lattice of Quantum Predictions", in: *International Journal of Theoretical Physics*, 32 (1993): 1853–1861; "Symmetry and Composition—a Key

M. Drieschner (ed.), *Carl Friedrich von Weizsäcker: Major Texts in Philosophy*, SpringerBriefs on Pioneers in Science and Practice 23,
DOI: 10.1007/978-3-319-03671-7,

to the Structure of Physical Logic?", in: *International Journal for Theoretical Physics*, 37 (1998): 427–733; "Reality, Viewed from Quantum Mechanics", in: H.D. Doebner et al. (eds.): *Trends in Quantum Mechanics.* (Singapore: World Scientific 2000): 86–95; (with Tim Oliver Eynck und Holger Lyre): "Comment on Redhead: The Interpretation of Gauge Symmetry", in: Kuhlmann M.; Lyre, H.; Wayne, A. (eds.): *Ontological Aspects of Quantum Field Theory* (Singapore: World Scientific, 2002): 303–312; "Is Time Directed?", in: Albeverio, Sergio; Blanchard, Philippe (eds.): *Direction of Time* (Heidelberg-New York etc.: Springer, 2014): 117–135.

Home page http://www.ruhr-uni-bochum.de/philosophy/staff/drieschner.

About the Book

This book presents a collection of texts by the German philosopher and physicist Carl Friedrich von Weizsäcker (1912–2007), for use in seminars on philosophy, mainly epistemology and the philosophy of physics or foundations of quantum mechanics, but also for courses on German philosophy of the twentieth century or the philosophy of science. Weizsäcker became famous through his works in physics, later becoming well known as a philosopher and an analyst of contemporary culture and politics. He worked intensively on projects for the prevention of nuclear war and for peace in general. Texts about classical, especially ancient philosophy are included as well as on foundations of science (especially of quantum physics), on logic, on the philosophy of biology and on power.

Besides an Introduction by the editor this volume includes ten major philosophical texts of Carl Friedrich von Weizsäcker: Preliminary Epistemological Considerations—A Description of Physics—Time–Physics–Metaphysics—Biological Preliminaries to Logic—Models of Health and Illness, Good and Evil, Truth and Falseness—Parmenides and the Graylag Goose—Parmenides and Quantum Theory—Possibility and Movement: A Note on Aristotelian Physics—The Rationality of Emotions—On Power.

M. Drieschner (ed.), *Carl Friedrich von Weizsäcker: Major Texts in Philosophy*, SpringerBriefs on Pioneers in Science and Practice 23,
DOI: 10.1007/978-3-319-03671-7, © The Author(s) 2014

MIX
Papier aus verantwortungsvollen Quellen
Paper from responsible sources
FSC® C105338

If you have any concerns about our products,
you can contact us on
ProductSafety@springernature.com

In case Publisher is established outside the EU,
the EU authorized representative is:
Springer Nature Customer Service Center GmbH
Europaplatz 3, 69115 Heidelberg, Germany

Printed by Libri Plureos GmbH
in Hamburg, Germany